HSE
Health & Safety
Executive

HEALTH AND SAFETY IN KITCHENS
& FOOD PREPARATION AREAS

GW00713262

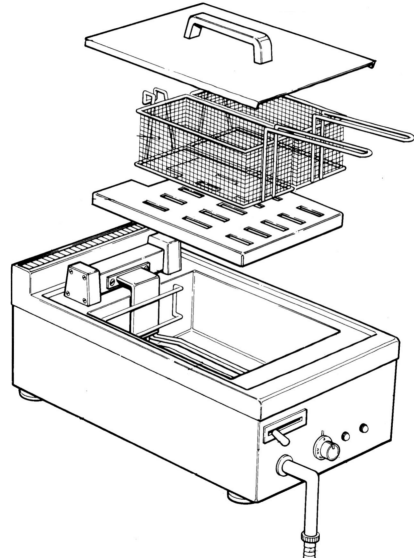

©Crown copyright 1990
First published 1990
Reprinted 1991
HS(G) series
ISBN 0 11 885427 5

The purpose of this series is to provide guidance for those who have duties under the Health and Safety at Work etc Act 1974 and other relevant legislation. This booklet gives advice on the hazards associated with equipment commonly used in catering and related industries and the precautions necessary to achieve a safe working environment.This guidance should not be regarded as an authoritative interpretation of the law.

The illustrations are to assist in understanding the text and are not intended to represent any particular manufacturer's product.

Further advice on this or any other HSE publications may be obtained from Area Offices of HSE or from the general enquiry points listed below.

Health and Safety Executive
Library and Information Services
Broad Lane
SHEFFIELD S3 7HQ
Telephone: (0742) 752539
Telex: 54556

Health and Safety Executive
Library and Information Services
St Hugh's House
Stanley Precinct
Trinity Road
BOOTLE
Merseyside L20 3QY
Telephone: 051-951 4381
Telex: 628235

Health and Safety Executive
Library and Information Services
Baynards House
1 Chepstow Place
Westbourne Grove
LONDON W2 4TF
Telephone: 071-221 0870
Telex: 25683

CONTENTS

London: HMSO

FOREWORD

This booklet provides guidance for those who have
duties under the Health and Safety at Work etc Act
1974 and other relevant legislation. It gives advice
on the hazards associated with equipment found
in catering and related industries and the
precautions necessary to achieve a safe working
environment. This guidance should not be
regarded as an authoritative interpretation of the
law.

The booklet is the third in a series giving guidance
to those in catering, food preparation and similar
activities. The other two booklets are HS(G)35,
Catering safety - Food preparation machinery, and
HS(G)45, *Safety in meat preparation - guidance for
butchers.*

The illustrations are to assist in understanding the
text and are not intended to represent any
particular manufacturer's product.

INTRODUCTION

1 Much of the equipment used in preparing, cooking and serving food in catering establishments, shops and food factories is so familiar that it is easy to forget the potential hazards it can present during use, cleaning and maintenance.

2 This booklet is intended for the owners and managers of catering establishments, shops and small food factories and also for employees and safety representatives. It describes the main hazards associated with the most common equipment used and the precautions that should be taken to safeguard those who work with it. It has been prepared by the Health and Safety Executive's Food and Packaging National Industry Group, and the Local Authority Unit, after wide consultation with employers, trades unions, equipment manufacturers and other interested organisations.

3 The Health and Safety Executive (HSE) is responsible for setting standards nationally, and for inspection and enforcement in food factories. Local authorities are responsible for inspection and enforcement in shops, hotels and catering establishments. Further information and advice, including detailed advice on legal requirements, may be obtained from Area Offices of HSE or the Local Authority Environmental Health Department.

4 The Hotel Catering Training Company recommends this booklet as valuable support material for use in training staff and suggests that it has particular relevance to those undertaking safe working modules in its workbase qualification scheme 'Caterbase'.

LEGAL DUTIES

5 Under the Health and Safety at Work etc Act 1974, employers have a general duty to ensure, so far as is reasonably practicable, the health, safety and welfare at work of their employees and, where appropriate, non-employees. This duty includes the provision of safe plant and equipment, a safe workplace, and the necessary information, instruction, training and supervision. In addition, employers who employ five or more people must have an effective health and safety policy which sets out the arrangements for fulfilling their duty, including how they intend to ensure that the necessary safeguards are adopted. Employers should consult safety representatives appointed by recognised trade unions.

6 Employees have a duty under the Act to take reasonable care for their own safety and the safety of others who may be affected by their actions: they should cooperate with their employer so far as it is necessary to enable their employer to comply with the Act.

7 Equipment suppliers must ensure that, so far as is reasonably practicable, equipment is designed and made so that it can be used safely. They must provide adequate information on safe operation, use and maintenance.

8 This booklet is not intended as a complete safety specification for suppliers or manufacturers. Users should seek advice from suppliers on any modifications which are practicable and available to reach the standards described.

9 Specific requirements on health, safety and welfare in the workplace are also laid down in:
- Factories Act 1961
- Offices, Shops and Railway Premises Act 1963
- Dangerous Machines (Training of Young Persons) Order 1954
- Prescribed Dangerous Machines Order 1964
- Fire Precautions Act 1971
- Fire Precautions (Factories, Offices, Shops and Railway Premises) Order 1976
- Health and Safety (First-Aid) Regulations 1981
- The Gas Safety (Installation and Use) Regulations 1984
- The Reporting of Injuries, Diseases and Dangerous Occurrences Regulations 1985
- The Control of Substances Hazardous to Health Regulations 1988
- Electricity at Work Regulations 1989
- The Pressure Systems and Transportable Gas Containers Regulations1989

STAFF TRAINING

10 Many accidents occur when employees use equipment without proper training. It is illegal for staff to use the equipment described in this booklet without proper training and, when necessary, supervision. The catering industry has many employees who change jobs frequently; it also uses a high proportion of seasonal and casual labour. This makes training even more important. Managers and supervisors may also need suitable training, particularly where they will be responsible for training staff to use hazardous equipment.

11 Supervisors and, whenever possible, staff should be present when equipment installers commission and demonstrate how to use new equipment. User handbooks which detail use,

working practices and cleaning routines should be retained for training purposes.

12 The following check list shows the main stages that should be considered in a typical training programme.

(a) Training checklist
Information and advice may be obtained from:
The supplier of the equipment (eg in the user handbook)
Training Centres and Colleges of Further Education
Trade Associations
Trades Unions

(b) Organisation
(i) How is the trainee to be selected?
(ii) Who is to supervise the training?
(iii) Who is to do the training?
(iv) What records of training will be kept?

(c) Assessing the trainee
(i)What is the trainee's existing knowledge?
(ii) Has the trainee worked with similar equipment elsewhere?
(iii)If trained elsewhere has the trainee an adequate knowledge of safe working practices?
(iv) Has the trainee special needs, for example, language difficulties?

(d) Basic instruction
(i) Prepare a check list of all the points that the trainee must remember. There may be a need for multi-lingual notices or easily understood graphics.
(ii) Explain how the equipment works.
(iii) Explain the dangers that can arise from the use and misuse of the equipment.
(iv) Explain the safety features of the equipment and how they protect the operator.
(v) Explain how to operate the equipment and how to follow any emergency procedures, including isolation from electrical, gas or steam supplies.
(vi) Explain how to clean the equipment safely. No one under 18 years of age should be allowed to clean machinery if there is a risk of injury from a moving part of that or any adjacent machinery.
(vii)Explain what to do if the equipment is or seems faulty.

(e) Supervised working
(i) Ensure that the supervisor has sufficient knowledge of all the equipment,processes, hazards and precautions and fully understands the responsibilities involved in training staff.
(ii) Set the trainee to work under close supervision, ideally following a prepared training programme.
(iii) Make sure the supervisor has sufficient time and knowledge to supervise effectively.
(iv) Make sure the supervisor watches for bad habits or dangerous practices that may develop, and stops them.

(f) Finally assessing the trainee
(i) Check that the trainee knows how to use and clean the equipment properly and safely.
(ii) Make sure that the trainee can be left to work safely without close supervision.
(iii) Complete and sign the training record.

FIRST AID

Regulations

13 The Health and Safety (First-Aid) Regulations 1981 apply to all workplaces. Guidance on the Regulations is given in HSE booklet COP 42, *First aid at work*.

First aid at work

14 How much first-aid equipment should be provided in a workplace depends on the number of people employed. For a small establishment a single first-aid box may be sufficient. This should be in the charge of a responsible person, and should be properly stocked. Advice can be obtained from the Local Authority Environmental Health Department, the Employment Nursing Adviser at your local HSE Area Office, or from a supplier of first-aid equipment.

Burns and scalds

15 Burns and scalds are common injuries in the catering industry. Burns are caused by, for example, touching hot surfaces or picking up a hot dish or utensil. Scalds are caused by hot liquids, such as boiling water, steam, or hot oil or fat. Chemical burns can be caused by some cleaning fluids.

16 The effects are pain, redness, swelling, blistering and often shock. In a kitchen someone should be available who knows how to deal with burn and scald injuries. A notice stating the action to take if someone is burned or scalded should be permanently displayed in the kitchen. Figure 1 shows suitable wording.

Burns and scalds - First aid

Immediately immerse the affected part in
(or pour over) cold running water for at least
10 minutes or until the pain is relieved.

Remove anything that may cause constriction
if there is swelling, eg rings, belts, shoes.

Cover affected area with a dry sterile dressing.

Send to hospital if serious.

Figure 1

REPORTING ACCIDENTS

17 Under the Reporting of Injuries, Diseases and Dangerous Occurrences Regulations 1985 (RIDDOR) employers have a legal duty to report certain accidents, dangerous occurrences and occupational diseases to the relevant enforcing authority (see paragraph 3).

Immediate notification

18 You must notify the appropriate authority as soon as possible, normally by telephone, if:
(a) someone dies, or suffers a major injury, in an accident in connection with your business;
(b) there is a dangerous occurrence (such as a burst steam boiler).

Report in writing

19 You must send a report to the appropriate authority within 7 days if:
(a) an employee is off work for more than 3 days as a result of an accident at work;
(b) you have previously notified by telephone any death, major injury or dangerous occurrence;
(c) a specified occupational disease is certified by a doctor.

20 Reports should be made on Form 2508, for accidents and dangerous occurrences (Figure 2 shows a reduced copy of Form 2508). Form 2508A should be used for reporting cases of disease. These forms are available from HMSO bookshops. Photocopies of the forms may be used.

Record keeping

21 You must keep a record of any reportable accident, dangerous occurrence or case of disease.

WORKING ENVIRONMENT: GENERAL POINTS

22 The following points apply whenever catering equipment is used.

Layout

23 The layout of the kitchen will depend upon its area and the items of equipment in it. There should always be enough room around equipment for staff to move around safely without bumping into each other.

24 Staff using knives and other hand tools should have enough room to work safely. There have been

Health and Safety Executive
Health and Safety at Work etc Act 1974
Reporting of Injuries,Diseases and Dangerous Occurrences Regulations 1985

Report of an injury or dangerous occurrence

- Full notes to help you complete this form are attached.
- This form is to be used to make a report to the enforcing authority under the requirements of Regulations 3 or 6.
- Completing and signing this form does not constitute an admission of liability of any kind, either by the person making the report or any other person.
- If more than one person was injured as a result of an accident, please complete a separate form for each person.

A   Subject of report *(tick appropriate box or boxes)* — *see note 2*

Fatality ☐ 1   Specified major injury or condition ☐ 2   "Over three day" injury ☐ 3   Dangerous occurrence ☐ 4   Flammable gas incident (fatality or major injury or condition) ☐ 5   Dangerous gas fitting ☐ 6

B   Person or organisation making report (ie person obliged to report under the Regulations) — *see note 3*

Name and address —

Nature of trade, business or undertaking —

If in construction industry, state the total number of your employees —

and indicate the role of your company on site *(tick box)* —

Main site contractor ☐ 7   Sub contractor ☐ 8   Other ☐ 9

Post code —

Name and telephone no. of person to contact —

If in farming, are you reporting an injury to a member of your family? *(tick box)*   ☐ Yes   ☐ No

---

C   Date, time and place of accident, dangerous occurrence or flammable gas incident — *see note 4*

Date ☐ ☐ 19   *day month year*   Time —

Give the name and address if different from above —

.ENV

Where on the premises or site —
and
Normal activity carried on there

Complete the following sections D, E, F & H if you have ticked boxes, 1, 2, 3 or 5 in Section A. Otherwise go straight to Sections G and H.

D   The injured person — *see note 5*

Full name and address —

Age ☐   Sex ☐ (M or F)   Status *(tick box)* —   Employee ☐ 10   Self employed ☐ 11   Trainee (YTS) ☐ 12   Trainee (other) ☐ 13   Any other person ☐ 14

Trade, occupation or job title —

Nature of injury or condition and the part of the body affected —

F2508 (rev 1/86)   *continued overleaf*

incidents when one member of staff has accidentally stabbed another because they were working in cramped conditions.

25 There must be room for staff to move trolleys and to carry trays safely. This is particularly important around equipment which has an exposed hot surface, for example a griddle top.

26 Side hinged doors, and bottom hinged doors that open just above floor level, should not obstruct a gangway.

27 Hazards can be created by placing some items of equipment next to others, for example a deep fat fryer next to a sink, or a shelf above an open top range.

28 To avoid staff bumping into each other there should wherever possible be clearly marked IN and OUT doors to kitchens and dining rooms etc. Where only one door is available it should contain a wired clear glass panel of adequate size to give sight of any person coming in the opposite direction. Your Local Environmental Health Department will give you advice on the requirements of the Food Hygiene Regulations.

Floors

29 Slips are the main cause of accidents in kitchens. A slip-resistant floor surface should be provided in kitchens, serveries and dining areas. Adequate drains should be provided in kitchens to carry away water, steam drips and waste from tilting kettles, brat pans and other similar equipment. The floors should be kept in good condition and clear of spilled fat, water, food, etc on which staff could slip. Spillage of any kind should be cleaned immediately. Manufacturers or installers will advise on the correct cleaning materials and methods for a particular floor surface. The use of incorrect cleaners could destroy the non-slip properties and may cause sheet flooring to lift at the seams. Hot pots and pans should not be placed on the floor. Protective footwear can also have slip-resistant properties.

Equipment stability

30 All catering equipment should be installed on a level surface on a secure base. Where castors are fitted the brakes should be regularly checked to make sure they are working properly. Smaller pieces of equipment that sit on a work top should be stable and positioned so that they cannot be dislodged.

Figure 2 Report Form 2508

Asbestos: removal and repair of equipment

31 Asbestos was used for insulation, as lagging, and in door seals in some catering equipment. It may still be present in older equipment. Lagging should be protected against damage where possible. Where it becomes damaged it should be resealed. Any insulation, lagging or door seal that is to be removed or repaired should be checked for asbestos before work starts, if it is not already known to be asbestos free. If asbestos is present, advice on the special precautions necessary should be obtained from your local HSE Area Office or Environmental Health Department before any work starts.

Lighting

32 Lighting must be good enough to enable employees to see the equipment and products properly and to clean effectively. Light fittings should be positioned so that there is an even light throughout the working area without glare or shadows. They should be cleaned regularly. Catering departments should be illuminated to the following standards:
(a) preparation and cutting rooms 540 lux
(b) kitchens 300 lux
(c) passages and store rooms 150 lux.

Temperature

33 Because of the very nature of the cooking process, and the need to serve cooked food hot, high temperatures and humidity are not unusual in kitchens and serveries. Both can affect the health, comfort and efficiency of kitchen staff. Ventilation, with sufficient air changes and adequate movement of air, is necessary to cool the workplace, and counteract humidity.

34 Fume extraction alone may not be adequate to ventilate properly all parts of the kitchen and, if necessary, the servery. Additional extractor or circulation fans may be necessary. Air inlets should be carefully sited to make sure that there is air movement in all parts. In kitchens where the temperature or humidity is persistently high the advice of a ventilation engineer should be sought.

Warning notices

35 Warning notices and approved safety signs can remind employees and others of danger and of safe practices. Many equipment suppliers supply suitable notices. The points to remember for each piece of equipment described in this booklet could form the basis for warning notices. Use warning notices to warn of temporary danger, such as a

wet floor after cleaning.

36   Check if warning notices need to be in any foreign language.

Cleaning

37   This booklet deals with the basic good cleaning practices for catering equipment which are fundamental to safety and good food hygiene. Your local Environmental Health Officer will advise you on other aspects of food hygiene practice. The booklet does not give advice on particular cleaning products. Specialist advice on cleaning methods and products can be obtained from suppliers of cleaning agents and equipment.

38   Catering equipment is usually cleaned before cooking begins or, preferably, at the end of the working day so that the equipment is left clean overnight. Staff may be tired at the end of their working day, however, and tempted to take shortcuts, so close supervision will be necessary .

39   The person cleaning the equipment may have to work near hot surfaces or empty hot liquids from the equipment. There is a risk of serious injury if safe methods of work are not clearly set down and followed. Supervisors and managers should make sure that such methods and procedures are established and monitored. Catering equipment should not be cleaned before the power supply is isolated*, or unplugged, or the gas supply, including pilot lights, is turned off. The equipment should be allowed to cool before cleaning starts.

40   Many cleaning materials and detergents are substances hazardous to health (for example, corrosive cleaning chemicals can cause serious skin burns), and will be subject to The Control of Substances Hazardous to Health Regulations 1988. This means that you must prevent, so far as reasonably practicable, staff being exposed to these substances, for instance by using a safe alternative if one is available. When you cannot prevent exposure you must adequately control it. This can often be achieved by:
• using automatic detergent dispensing equipment;
• buying diluted substances so that staff do not have to handle concentrated substances;
• using non-spill and unbreakable containers;
• transporting, transferring and storing substances in a safe way, for example in small quantities;
• after taking all reasonably practicable precautions to prevent or adequately control exposure as far as possible by other means,

*see paragraph 44

providing staff with suitable personal protective clothing, for example rubber gloves and eye protectors.

41   Make sure the cleaning agent is suitable for the equipment. Follow the equipment manufacturer's recommendations. Never transfer cleaning chemicals into other containers such as flour shakers, milk or soft drink bottles. This can lead to confusion and serious accidents. Never mix cleaning chemicals together. Some contain chemicals based on chlorine and some contain acid; if mixed, they give off toxic chlorine gas. Cleaning chemicals should be stored in a separate clearly marked area away from food or food preparation equipment. Dilute cleaning chemicals in accordance with the supplier's instructions.

42   Most catering equipment uses electricity. If water gets into electrical equipment it can cause electric shock or a dangerous malfunction. The risk of water getting into electrical equipment is greater where pressure washers, steam cleaners or hoses are used. Never clean equipment with these unless the electric wiring and circuits are specially protected against water; the equipment manufacturer will be able to advise you. Wet washing walls can result in water seeping into light switches and socket outlets. Where this could happen suitable waterproof connectors and fittings should be installed.

Cleaning check list

43   Follow the laid down cleaning procedure or the advice in the equipment manufacturer's users' handbook. Safe cleaning methods should be simply and clearly set out and anyone who cleans the equipment should fully understand them.

44   Before starting to clean or remove any part of the equipment, switch off and isolate the electrical supply or turn off the gas or steam supply as appropriate. The isolator is the main switch which cuts off the electrical supply to the equipment. Do not rely solely on the equipment controls. Where the isolator is located some distance away from the equipment being cleaned, it should be locked in the OFF position with the keys in the possession of the person at risk, and a suitable warning notice should be hung over the isolator. Where the equipment is connected to the electrical supply by a plug and socket, always unplug it before cleaning starts.

45   Follow the instructions supplied with the equipment and cleaning chemicals. Always use the recommended protective clothing (including gloves, aprons, eye protection, boots, etc).

46 Make sure water does not get into any electrical equipment, wiring or controls.

47 After cleaning do not operate the equipment until it has been properly reassembled and checked.

ELECTRICAL SAFETY

48 The Electricity at Work Regulations 1989 impose duties on employers, the self-employed and employees to take precautions against the risk of death or personal injury from electricity in work activities. Electricity at normal mains voltage (240 V) can cause fatal shock, burns and fire. Wet conditions increase the risk of electric shock, so particular care is needed in catering and food preparation premises.

49 All electrical equipment should be properly installed, serviced and maintained by a qualified electrician. Untrained people can easily make deadly mistakes, putting themselves and other people at risk, and should not carry out any electrical work.

50 Each piece of equipment supplied through a permanent cable must have its own isolator* or plug and socket arrangement so that it can be disconnected from the electrical supply for cleaning or repair. Each isolator and dedicated socket outlet should be clearly labelled to show which equipment it supplies. Isolating switches should preferably have a facility for locking them in the OFF position.

51 Mobile and smaller pieces of equipment may be connected to socket outlets by flexible cables. Industrial type plugs and sockets (to British Standard 4343) are preferable to the domestic type which are often unsuitable for commercial catering equipment. Working areas should have sufficient socket outlets to avoid the use of multiple socket adaptors, extension leads or training cables.

52 Efficient cable or cord grips should be used both at the plug and where the cable enters the equipment. Use the correct fuse. The earth wire (where provided) must always be properly connected. Loose, cracked or broken plugs should be taken out of use immediately and the damaged part replaced.

In this booklet the term 'isolator' describes any means of effective disconnection of the supply. It includes switch fuses, switches and isolators used as part of the fixed wiring installation.

53 Flexible cables should be positioned and protected so that they cannot be easily damaged. They should not trail across hot or heated surfaces. They should be checked regularly for damage and loose connections. Cables to equipment in every day use should be checked at least once a week. Some cables can be seriously affected by animal fats, oils and cleaning fluids. If a cable is damaged, or shows signs of swelling or cracking, the equipment should be taken out of use and the cable replaced. Do not carry out make-shift repairs to damaged cables.

54 There is an increased risk of electric shock if water gets into electrical equipment. Do not let water get into any electrical equipment during cleaning. Hoses and pressure washers create the greatest risk: do not use a hose to clean equipment that is not suitably constructed†. Socket outlets should not be sited where they can get wet. Domestic 13-amp square-pin plugs are not suitable for use in wet or moist conditions. If such conditions are likely splashproof, hoseproof or watertight electrical equipment should be used.

55 A sensitive (30-mA max) residual current device (RCD) (also known as an earth leakage circuit breaker) should be fitted in the supply to pressure washing units and steam cleaners. These devices can appear to be working when they are not, so if fitted they must be checked regularly by means of the test button provided. They are in addition to and not a substitute for proper installation and maintenance of the whole electrical system.

56 START buttons should be recessed or shrouded to prevent unintended operation.

57 STOP buttons should be red, protruding for easy operation and within easy reach of the operator.

58 Control functions should be clearly marked.

59 If in doubt, ask the advice of a competent electrician.

†If the equipment is marked (usually on the name plate) with a code starting with the letters IP, it is an indication of its ability to withstand the ingress of dust and water. If the second number of the code is 4 (eg IP44 or IP 54) the equipment is protected only against splashing. If the second number of the code is 5 (eg IP 65) the equipment is protected against low pressure jets. If 6 is the second number the equipment has an even higher degree of protection.

GAS SAFETY

62 Gas, including liquefied petroleum gas (LPG), is widely used in the catering industry as a source of direct heat for ovens, boiling tops, grillers etc and also for heating water in steam boilers, water sets etc.

Hazards

63 The main hazards associated with gas are:
(a) fire and possibly explosion when accumulations of unburnt gas are ignited; and
(b) carbon monoxide poisoning from gas which is not burned properly. Carbon monoxide is odourless and tasteless and therefore hard to detect. It can be given off by installations which are faulty or inadequately maintained. It is highly poisonous and breathing it can quickly lead to death.

Precautions

64 Gas appliances should be installed in accordance with the relevant British Standard BS6173, *Installation of gas catering appliances.*

65 Each gas appliance should be installed in a well lit and draught free position. Ventilation, whether natural or mechanical, should be provided to ensure an adequate supply of fresh air, otherwise the gas will not burn completely and poisonous carbon monoxide will be produced. The outlets should never be covered or added to, and air inlets should be kept free of obstruction.

66 Gas appliances should be installed, fitted and maintained only by a competent person trained in accordance with the Approved Code of Practice *Standards of training in safe gas installation* such as employed by installers registered with the Confederation for the Registration of Gas Installers (CORGI). The Gas Safety (Installation and Use) Regulations 1984 are concerned with users' safety and under the regulations persons must not install or service any gas supply, appliance or flue if they are not competent in gas installation and servicing.

67 Gas appliances should be regularly serviced by a competent gas service engineer. The appliance manufacturer's instructions should say what the user should do in this connection and how often. Always follow the manufacturer's instructions.

68 It is common practice to install a gas shut off valve in the kitchen to shut off the gas supply to all the appliances in the kitchen in an emergency. Staff should know where this gas valve is situated, or where the main gas valve at the meter is situated, so that in an emergency they can turn off the gas supply to the kitchen. The appliance gas control taps should be turned off at the end of each working period.

69 In addition to the appliance gas control taps there should be a gas shut off valve installed in an accessible position close to the appliance to allow gas to be shut off for routine maintenance or in an emergency.

70 If the gas has been turned off at a main gas valve in the kitchen, or at the meter, only a trained member of staff should relight the appliances or pilot lights after the gas is turned back on.

71 Ignition jets and pilot lights should be kept clean and regularly serviced.

72 If an integral ignition device fails repeatedly to ignite the gas it should be reported to the supervisor.

73 Appliances with manual ignition should always be relit with a taper.

Gas leaks

74 If you smell gas:
(a) do not use any naked lights;
(b) do not switch the lights or any other electrical equipment on or off: switches produce sparks that would ignite escaping gas;
(c) check whether gas is coming from a pilot or burner:
    (i) if so, turn off the burner;
    (ii) if not, turn off the supply where it enters the room or at the meter;
(d) open doors and windows to get rid of the gas and leave them open until the leak has been stopped and any build up of gas dispersed;
(e) report the leak immediately to the person in charge;
(f) do not turn the gas back on where it enters the room or at the meter until the fault has been traced and repaired by a competent gas service engineer;
(g) if gas continues to escape after the supply has been turned off at the meter, contact the Gas Suppliers Emergency Service immediately. The number to call is on a label on or near to the meter.

Liquefied petroleum gas

75 Many small catering establishments use liquefied petroleum gas (LPG) as a fuel supplied either from fixed (bulk) storage tanks or from

cylinders. Unlike mains gas, LPG is heavier than air. Users of LPG should observe the following guidance in addition to the advice given on gas safety.

76 Storage of LPG at fixed installations in tanks refilled on site from road tankers should comply with HSE booklet HS(G)34 which gives guidance on LPG tanks including safe location, protection against damage from vehicles, security against vandalism, electrical equipment and provision of fire fighting equipment. Further extensive guidance is also published by the Liquefied Petroleum Gas Industry Technical Association (UK) (LPGITA).

77 LPG cylinders should be stored in compliance with the advice given in HSE booklet HS(G)34 and the Approved Code of Practice *Safety of transportable gas containers*. It is preferable that they be stored in the open air.

78 Make sure the storage areas for LPG whether in bulk storage tanks or cylinders are kept free from weeds, grass and other combustible materials and that cylinders do not normally stand or lie in water.

79 Installation work, servicing and maintenance should be carried out only by a competent person.

80 Wherever possible LPG appliances conforming to a British Standard should be used. This will ensure that it is a properly tested appliance design incorporating appropriate safety controls. The appliances should be regularly serviced and maintained by competent service engineers. If any difficulty or problem occurs advice should always be sought from the LPG gas or appliance supplier.

81 Only properly trained staff should change a gas cylinder: when changing cylinders the following precautions should be followed:
(a) handle all cylinders with care. They should not be dropped or allowed to come into violent contact with any other object;
(b) handle and use cylinders in an upright position - outlet valve uppermost;
(c) extinguish any fire, flame, cigarette or source of ignition including pilot lights;
(d) turn off all appliance gas taps;
(e) make sure that the replacement gas cylinder is the correct one for the installation;
(f) make sure that the replacement cylinder connections are gas tight. Any leaking gas will smell; if necessary test by brushing with soapy water around the connection; bubbles will form if gas is leaking. Never use a naked flame;

(g) make sure that the gas cylinder valve is closed before you disconnect the empty cylinder, or before you remove any plastic cap or plug on the outlet connection of the replacement cylinder;
(h) make sure that the regulator tap is closed before you disconnect cylinders with self sealing valves;
(i) always use the correct spanner on connections designed to be tightened with a spanner; hand tightness is not sufficient.

STEAM SAFETY

82 Steam-heated catering equipment includes steaming ovens, bulk boiling pans, bains- marie, hot cupboards, steam cupboards, water boilers and some beverage machines.

83 Steam may be generated within the equipment, or be supplied to the equipment from a separate steam boiler (sometimes called a steam generator): in this case there will be associated lagged pipework which should include on the outlet side of the boiler a steam trap and a pressure reducing valve. The equipment will always be pressurised by the steam, albeit at low pressure. Independent small steam boilers are used for making hot drinks in serving areas. The Pressure Systems and Transportable Gas Container Regulations 1989 will apply to this type of equipment.

Hazards

84 The main hazards associated with steam-heated equipment are boiler explosion due to over-pressurisation or lack of water, explosion of the equipment due to over-pressurisation, and scalding, often caused by hot water and steam escaping when the equipment door is opened. Fittings in the steam supply, such as valves, can be very hot.

Precautions

85 Steam boilers should have the following safety fittings: safety valve, pressure gauge, water level gauge (glass), low water level cut-out device, blow down valve and shut-off valve. The boiler and its fittings should be thoroughly examined at least once every twelve months by a competent person, normally an engineering surveyor from one of the inspection authorities employed by insurance companies.

86 Steam boilers, water inlet valves, pipes and tanks should be regularly cleaned and any scale

removed: this is particularly important in hard water areas. The steam trap should be regularly drained and properly maintained. This work should be done only by a service or maintenance engineer.

87   A steam boiler should be situated so that the pressure gauge and water level gauge are easily and clearly seen: it should not be switched on if it does not contain sufficient water and when in use should be switched off if the water level gauge shows little or no water in the boiler.

88   Steam safety valves and blow-down valves should be vented to a safe place. Steam vents from steam-heated equipment should also vent to a safe place and must not be obstructed. Equipment which is pressurised should not be opened until the pressure is released. Staff should open the doors or lids of steam-heated cooking equipment carefully to avoid the escaping steam and water condensate.

Training

89   Staff who work with steam boilers or steam-heated equipment should be properly trained to use and clean the equipment. They should know why the safety fittings are there and what can happen if they do not work. They should be properly trained in the procedures to follow in an emergency.

Cleaning

90   Staff should never remove for cleaning the safety fittings on steam boilers and steam- heated equipment. Before cleaning, steam-heated equipment should be isolated from the steam supply, and allowed to cool. Staff should always follow the equipment manufacturer's users' instructions for cleaning.

91   *Remember*
- Check that the water supply to the steam boiler is turned on.
- Check that there is sufficient water in the steam boiler before you light the gas or switch on.
- Check the steam pressure regularly.
- Switch off the steam boiler if the pressure rises above the safe level.
- Switch off the steam boiler if the water level gauge shows insufficient water.
- If any steam comes from the safety valve during cooking, shut off the steam supply or heat and report immediately to the supervisor.
- When cooking is completed shut off the steam valve before opening doors and lids.
- Open doors and lids carefully and stand to one side to avoid contact with escaping steam.

- Wait until the equipment has cooled before starting to clean it.

HARD WATER - EQUIPMENT MAINTENANCE

92   Scale in water supply pipes, steam boilers, valves and equipment has caused serious accidents, including explosions. Scale builds up quickly in hard water areas. It restricts the aperture in pipes and valves and causes the water supply to be reduced or even blocked completely, which can lead to the equipment overheating. In hard water areas, catering equipment with water circulating inside should be regularly inspected and cleaned to remove scale. Even if a water softener is used, equipment should still be regularly inspected to check that the softener is effective.

93   How often cleaning is necessary will vary and is best judged by experience. Never neglect descaling, however, as build-up of scale could have catastrophic results.

94   Soft scale can be removed with a nylon brush and a good supply of hot water. Hard scale may require specialist treatment to remove it and this should be done carefully to avoid damaging the equipment. Never use sharp-edged tools or scrapers. Descaling with acids should be carried out only by properly trained staff using appropriate protective equipment. To prevent blockages water valves should be disconnected and cleaned regularly.

95   There should be easy access to, and removal of, the heating elements. For cleaning purposes removable inspection plates should provide access into water tanks.

FUME VENTILATION

96   Cooking fume ventilation hoods (or canopies) and ducting are needed to remove the smells, vapour and grease invariably produced in large quantities during cooking. The ventilation system should include a fan of sufficient extract capacity to cope with the expected fume load from the equipment which it serves, and be acoustically designed to reduce to a minimum the fan noise.

Hazards

97   The main hazard associated with fume ventilation equipment is fire caused by the ignition of accumulated grease and fat in the hood and the associated ducting.

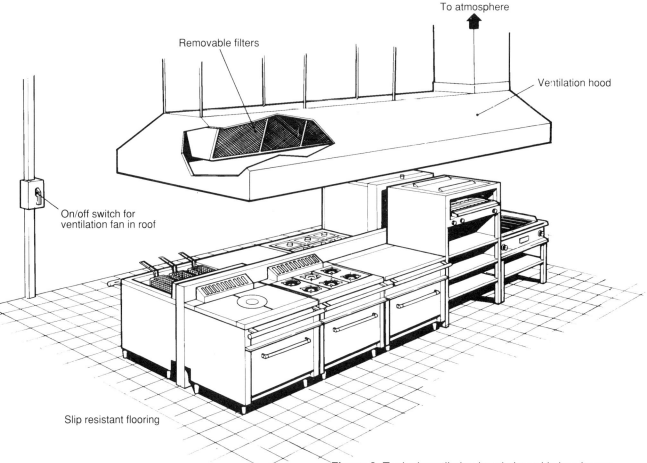

To atmosphere

Removable filters

Ventilation hood

On/off switch for
ventilation fan in roof

Slip resistant flooring

Figure 3 Typical ventilation hood above kitchen layout

Precautions

98 Ducting should be as short as possible,
preferably with the minimum number of bends, and
be vented directly to the open air. A grease filter
should be installed in the ventilation hood in a
readily accessible position within or as close to the
equipment as possible. If low level extract ducting
is installed a grease trap (sump) should be
provided.

99 Brick chimneys or flues and corrugated
ducting should not be used to conduct cooking
fumes unless they have been specially lined.
Chimneys or flues used to conduct fumes away
from catering equipment should have adequate
access for cleaning.

100 The ducting should serve only the kitchen
with no communication with the rest of the
premises. It should be separate from ventilation
required in the preparation and servery areas.
Where the ducting does have to pass through
other parts of the premises it should be contained
in a separate outer duct which has the same
standard of fire resistance as the kitchen, or the
parts of the building through which the duct
passes, if this is higher.

101 Clean cooker surfaces and hoods, and
empty and clean oil and condensation channels
daily. Remove and clean filters regularly, where
there is heavy use, a spare set should be
available. Clean automatic fire extinguishing
installations and automatic shutters or lids on
deep fat fryers daily.

102 Clean the inside surfaces of ducting, and the
fan blades, at least once every three months.
Before you clean ducting, switch the fan off and
allow sufficient cooling time.

103 Never hang combustible articles such as
clothes, towels and cloths over or near cooking
equipment with a fume ventilation hood.

104 Keep available an adequate supply of
appropriate firefighting equipment with clear
instructions for its use. Train employees to use it
correctly and, if there is a fire, to follow a laid
down procedure. Your local fire prevention officer
can give you advice on equipment, procedures
and staff training.

Training

105 Only trained staff, using a safe means of
access where necessary, should clean grease and
oil from hoods, fume ducts and extraction
equipment. The training should stress the potential
seriousness of fires in ventilation ductwork, and
how to use correctly the fire fighting equipment
provided.

106 *Remember*
•   Do not hang oven gloves, towels etc over or
     near cooking ranges, fryers or other hot
     appliances to dry.
•   If a fire starts carry out the laid down
     emergency procedure immediately.

SAFE USE OF KNIVES

107   Caterers use a range of different knives for a variety of tasks, for example cutting, slicing and dicing.

Hazards

108   Knife accidents are common in the catering industry. They usually involve cuts to the non-knife hand and fingers.

109   The food being cut is held in the non-knife hand and the knife is pushed down through the food. The work is often done at high speed and there is always a danger of cuts to the non-knife hand.

110   Cleavers are commonly used for chopping. The risk is the same as for knives but the injury can be more serious, even amputation of fingers.

Precautions

111   Select the correct knife for the task. Kitchen knives are generally designed for a particular job. Use only good quality kitchen knives.

112   Knives should be kept in good condition. They should be kept sharp and have handles that can be properly held. The handles should be kept clean.

113   When using a knife, use a firm grip, try to use even pressure for cutting, cut downwards and avoid cutting towards the body. Never try to catch a falling knife.

114   There should be enough room for there to be no danger of a person using a knife being bumped by another member of staff.

115   Cutting blocks, tables and boards should be firm, smooth and kept clean.

116   Knives should not be left lying about on worktops and tables. They should not be placed unprotected in cupboards or drawers, nor left in washing up water. They should be stored in suitable racks or sheaths.

117   When carrying knives hold the knife point downwards.

118   If a significant proportion of time is spent on cutting, particularly meat, a protective gauntlet should be worn on the non-knife hand. Advice on the precautions necessary when cutting meat, including deboning, is given in HSE booklet

HS(G) 45, *Safety in meat preparation: guidance for butchers.*

Training

119   Only trained staff should use kitchen knives. Staff should be trained to use the correct knife for the job, and the correct sharpening procedure.

120   *Remember*
*   Always use a knife suitable for the task.
*   Keep knives sharp.
*   Always hold the knife firmly.
*   Do not cut towards your body.
*   Do not leave knives on tables or in washing up water.
*   Put the knife away after use.
*   Always carry a knife point down.
*   Never try to catch a falling knife.

OVENS/RANGES

121   'General purpose' or conventional ovens rely on naturally circulating hot air within the cooking chamber.
*   Forced air convection (FAC) or convection or fan assisted ovens have a fan inside   which distributes heat uniformly throughout the interior.
*   Combination ovens are ovens that bake, roast, braise, grill or steam at atmospheric or low pressure. A potable water supply and drainage will be required for this type of oven.
*   Both forced air convection and combination ovens may have both gas and electric supplies.
*   Ovens can be free-standing or form part of a suite. A range is a composite unit made up of an oven underneath and a boiling table on top.

Hazards

122   The main hazard from ovens and ranges is being burned, either by touching a hot surface, or by being in the way of hot air when an oven door is opened. Ovens with bottom hinged doors can tilt forward if heavy meat joints are placed on the open door. Staff leaning over and cleaning behind a working oven risk burns from the flue.

123   There is a danger of a gas flashback if a gas oven does not light immediately or when gas which has built up in the oven because of flame failure is ignited.

124   Faulty electric wiring or broken door switches

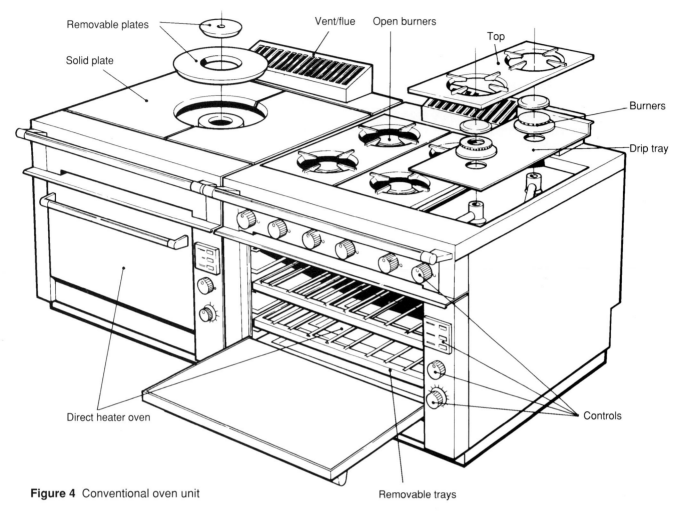

Removable plates

Solid plate

Vent/flue

Open burners

Top

Burners

Drip tray

Direct heater oven

Controls

Figure 4 Conventional oven unit

Removable trays

Figure 5 Fan assisted oven

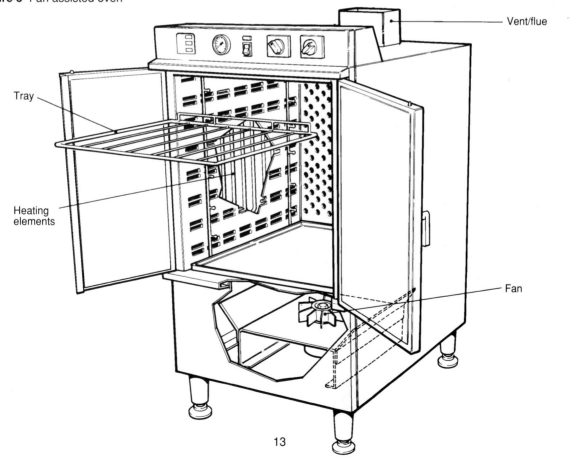

Vent/flue

Tray

Heating elements

Fan

13

on electric ovens can cause electric shock.

125   There is a possibility of injury from contact with moving unguarded fan blades in a forced convection oven if the fan is running during cleaning.

Precautions

126   A special oven cloth or oven gloves should be used when handling trays or tins in the oven. Similar care should be taken when moving oven racks or utensils on the hotplate or cooking top. Use only dry oven cloths.

127   Grills and salamanders should not be positioned over solid or open-topped ranges. Where a shelf is required over a cooking range, for example to hold service plates etc, the manufacturer's recommended fittings should be installed. Do not touch the metal sides of adjacent equipment as they can get very hot.

128   There should be sufficient room in front of an oven for an operator to stand clear when opening the oven door. Free standing ovens with bottom hinged doors should be stable.

129   The handles of saucepans should be placed away from the hotplate or gas ring, and not allowed to project beyond the edge of the range. Ladles or spoons should not be left in saucepans on hotplates or rings.

130   A safe lighting procedure for gas appliances should be carefully followed. Consult the users' handbook. Particular attention should be paid to the instructions on what to do if the burner/pilot fails to light. With manually lit ovens a lighted taper should be inserted before the gas supply is turned on. Make sure all the gas burners ignite. Pots and pans, serving dishes and other utensils should not be placed against or over air vents or flue outlets in the sides or top of an oven. Flue outlets should never be blocked.

131   Door-operated switches should be checked regularly to ensure they work properly. The fan in some forced convection ovens is switched off by a door-operated switch when the door is opened. It does not stop rotating immediately. The guard should always be in position before the oven is used.

Training

132   Staff who work with ovens or ranges should be properly trained in their use. They should know what to do if the gas fails to ignite immediately.

The purpose of door- operated switches should be explained.

Cleaning

133   Before an oven or range is cleaned, it should be switched off and isolated, or the gas turned off, and allowed to cool. Trained staff should clean the equipment following laid down procedures or the manufacturer's instructions. The fan guard should always be replaced after cleaning. Oven cleaning fluids and pads may contain caustic chemicals which can cause severe burns. Suitable gloves and eye protection should therefore be worn when they are used.

134   *Remember*
- Always use a dry oven cloth or oven gloves when handling anything in an oven or on a range.
- Metal surfaces of adjacent equipment may also be very hot.
- Always stand to one side when opening an oven door, and open the door slowly.
- Make sure the taper is lit before you turn on the gas supply.
- Make sure all the gas burners light and remain alight.
- If the fan interlock, if fitted, does not switch off the fan when you open the door report it to your supervisor.
- Do not leave bottom hinged oven doors open.
- Never use a forced convection oven if the fanguard is not in place.
- Do not rest anything heavy, for example a large meat joint while basting, on bottom hinged doors.

STEAMING OVENS

135   There are two types of steaming oven:
(a)   inject steam type - these are supplied with steam from an outside source, are pressurised and have a door interlock which prevents the door being opened until the pressure is equalised;
(b)   well type - these generate their own steam from an open tank of water sited under the cooking compartment. They can be heated by gas, electricity, or steam coil.

Hazards

136   The hazards associated with steaming ovens are scalds, particularly when opening the door, and burns from touching the hot outer casing.

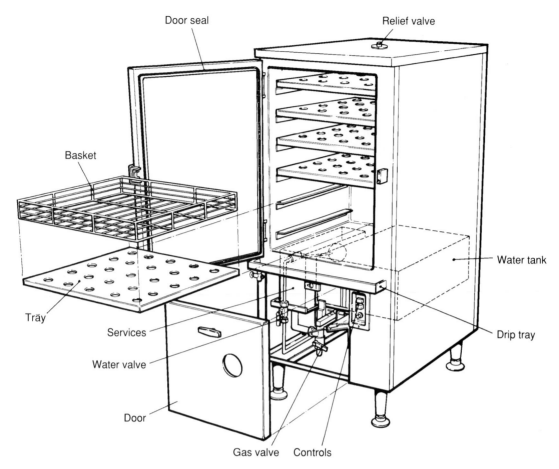

Door seal Relief valve

Basket

Water tank

Tray

Services

Drip tray

Water valve

Door

Gas valve Controls

Figure 6 Steaming cabinet

Precautions

137 Before turning on a steamer:
(a) check that the steam vent is open and unobstructed;
(b) check that the water supply tap is turned on, the brake tank is filled and there is water in the base of the oven. For electric steaming ovens check that there is sufficient water to cover the heating elements. The ball float regulating the water level should operate freely.

138 Close the steamer door firmly but without undue force and turn the door locking handle or wheel sufficiently to prevent steam leaking. Overtightening the door handle or unnecessarily slamming the door packs down the door seal gasket and shortens its life. Examine the door seal gasket periodically to check its serviceability.

139 The steamer should be sited where there is room for staff to stand well clear when the door is opened. When cooking is finished open the door carefully, partially at first to release the first rush of steam, and then fully. Stand well clear to avoid contact with the residual steam. Make sure the drip tray below the door is in position to catch water condensate as it drips from the inside of the door.

Training

140 Staff should be properly trained in the correct

methods of operating a steaming oven, the hazards involved and the precautions necessary.

Cleaning

141 Before cleaning starts the steamer should be turned off, and isolated if electric, and allowed to cool. The water supply should be shut off and the sump drained. Racks and their supports should be removed and the interior washed down, rinsed, dried and reassembled. Any scale should be removed with a proprietary descaler. The door should be left ajar to prevent noxious smells forming. Only trained staff should clean the steamer following a laid down procedure or the manufacturer's users' handbook.

142 *Remember*
• Before using the oven check that:
  (i) the electric heating elements are completely covered with water or that the sump is filled to the correct level;
  (ii) the ball float regulating the water level operates freely.
• Do not use the steamer if the steam vent is obstructed or not working.
• Open the door just a little at first to let the excess steam escape.
• Report any steam leakage around the door seal to your supervisor.
• Keep the drip tray in position.
• Clean daily or after use.

MICROWAVE OVENS

Hazards

143 The main hazards associated with using microwave ovens are burns or scalds caused when sealed containers containing hot food burst open. Hot food containers and steam also cause burns. Microwave ovens can catch fire if they are not used properly or if their contents overheat. Poorly sited ovens can cause the user back strain. Microwave energy could burn the user if the door seals are not effective or the protective mesh behind the glass door panel slips.

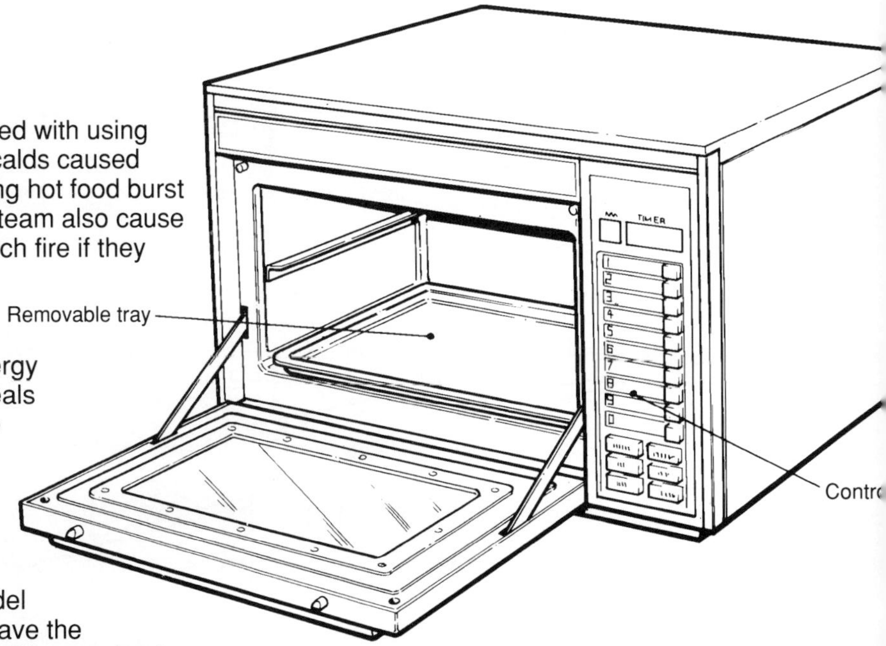

Removable tray

Precautions

144 Do not use a domestic model oven for commercial catering. Have the microwave oven regularly serviced by a trained engineer. Do not take the back off a microwave oven.

Figure 7 Microwave oven

145 The single most important precaution is not to put food in a sealed container in the oven unless the food manufacturer's instructions are to do so. A dish covered with, for example, unpierced clingfilm, or the shell round an egg, has the same effect in a microwave as a sealed container: either can burst open. Remove lids from jars and take-away food containers.

146 Food must not be cooked in metal containers or on metal plates unless they were supplied with the oven or the oven manufacturer says this is safe. Metal trays can be used in some microwave combination ovens but if in doubt, leave out. Do not use containers or crockery with metal decorations such as gold bands.

147 Use only clingfilm recommended for use in microwave ovens and puncture after covering the food products.

148 Do not cook food for longer than necessary. Take care when setting the timer.

149 When cooking foods with a high sugar or fat content, for example mince pies and Christmas puddings, follow the cooking instructions carefully.

150 Each day after use remove all traces of encrusted food, carbonised food and other foreign matter by cleaning the inside of the oven, the inside of the door and, if fitted, shelves and supports. The roof of the oven should also be

carefully wiped: take care not to damage the stirrer, if fitted.

151 Do not obstruct the air vents at the side and rear of the microwave. The oven should not be placed against a wall in a way that could obstruct the vents. The filters should be removed at least once a week, washed in warm soapy water, rinsed, squeezed dry and replaced.

152 The door should move freely and when closed seal the oven. The interlock switches on the door should switch off the oven as soon as the door is opened. Do not use the oven under any circumstances if the door does not close properly or the door interlock switches are broken.

153 Do not place the microwave under a counter or on a high shelf where loading and unloading food causes the operator unnecessary bending or stretching.

Training

154 Staff should be properly trained to use a microwave oven. The need for the door seal to be kept clean and the door closing mechanism to work properly should be stressed. If the oven is available for customer to use, clear operating instructions must be posted alongside.

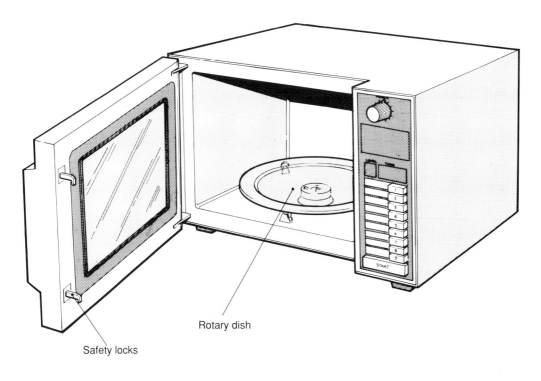

Rotary dish

Safety locks

Figure 8 Microwave with rotary dish

Figure 9 Back of microwave

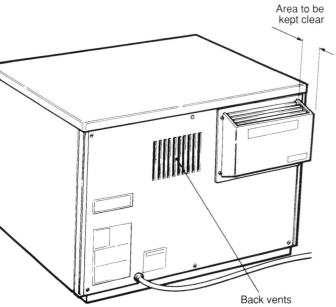

Area to be
kept clear

Back vents

Cleaning

155   All internal surfaces of the microwave oven, including the door, should be cleaned thoroughly every day after use. Pay particular attention to the door seals which should completely seal the oven when the door is closed. Remove any turntable, tray or supports and clean.

156   *Remember*
*   Do not put metal-decorated dishes in the microwave.
*   Do not put metal dishes in the microwave unless the manufacturer's users' handbook says this is safe.
*   Do not cook eggs in their shells in the microwave.
*   Always pierce clingfilm covering food in dishes before cooking.
*   Remove lids from jars and take-away food containers before you put them in the microwave.
*   Do not cook food in a sealed container unless it has been specially manufactured for use in a microwave oven.
*   Keep the inside surfaces of the oven and door clean.
*   Do not use the oven if the door does not close properly.
*   Take care when setting the time switch.
*   Do not use the oven if it does not switch off automatically when the door is opened.
*   Food containers can be very hot - use an oven cloth or oven gloves.
*   Remove clingfilm carefully and keep out of the way of the steam.
*   Never switch the oven on if it is empty.

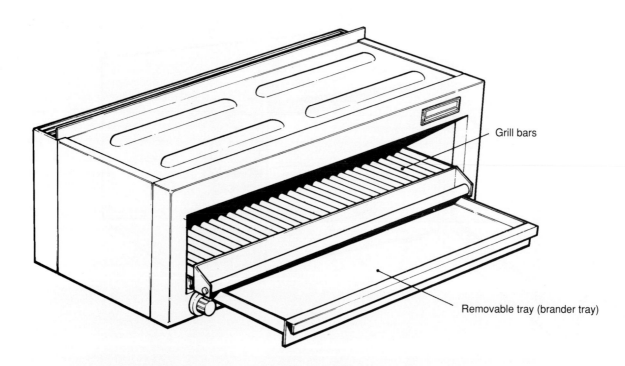

Figure 10 Salamander

GRILLERS

157   There are three basic types of griller: griddles, grillers and salamanders. They may be heated by gas or electricity.

Hazards

158   The main hazard associated with this equipment is burns from the gas burners or heaters, from hot trays and food or from a gas flashback. There is also a danger of the brander plate or shelf unit being pulled out, falling and injuring fingers or toes.

Precautions

159   Wall mounted units should be secured firmly in position. Do not position over a range top.

160   These appliances become extremely hot in use. Always use oven gloves or an oven cloth to handle cooking trays. Use tongs to handle hot food.

161   Grilling equipment is often at eye level. Staff should take special care not to burn their face and eyes.

162   There should be a minimum distance of 60 cm between the top of a grill and any overlying shelf or ceiling.

163   Follow the manufacturer's recommended lighting procedure for gas-heated grillers. If the automatic ignition, where fitted, fails, a lighted taper should be used to light the burners. Check regularly that the gas pilot lights are lit. Clean the gas burners regularly to prevent carbonised fat and food debris blocking them. If the burners do get blocked they should be cleared by a gas service engineer. Unskilled cleaning can 'open up' the gas jets and disturb the designed combustion rate.

Training

164   Staff who use grillers should be trained to use them safely. They should know the risks and the precautions to take.

Cleaning

165   Only trained staff should clean grillers. The equipment should be turned off and, if electric, isolated. The manufacturer's users' handbook recommended procedures should be followed. Any build-up of carbonised fat should be removed and specialised heavy duty cleaning may be required. Gas burner jets require regular maintenance to prevent blockage by food debris or carbon from whatever source. Remove the scrap drawer from griddles and clean. Brander plates may need specialised cleaning.

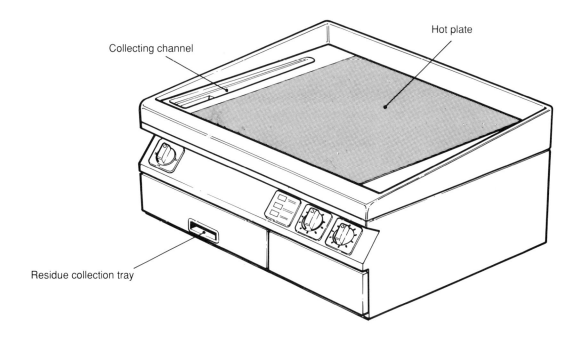

Collecting channel

Hot plate

Residue collection tray

Figure 11 Griddle

Figure 12 Griller

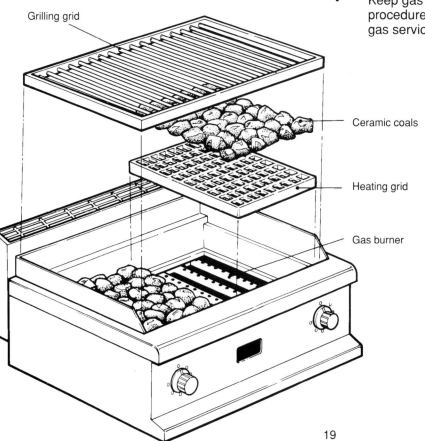

Grilling grid

Ceramic coals

Heating grid

Gas burner

166 *Remember*
- Check that all the gas burners have lit.
- Take special care to avoid burns to the eyes and face.
- Use an oven cloth or oven gloves to handle hot trays.
- Clean and remove carbonised fat regularly.
- Keep gas jets clear by normal cleaning procedures. If they do get blocked, call a gas service engineer.

DEEP FAT FRYERS

167 Deep fat fryers are used for cooking food quickly and efficiently. They may be heated by gas or electricity. They range in size from small table top single or double pan models to large multipan fixed units.

Hazards

168 The main hazard associated with deep fat fryers is burns from contact with hot cooking oil or fat. Burns can be ·caused if the hot oil or fat splashes when food or the basket is dropped in carelessly, or if it spits or boils over if there is excess water or moisture in the food. Fire from ignition of hot cooking oil or fat is also a major hazard. Spilled or splashed oil or fat on the floor around a fryer is a major slipping hazard.

Precautions

169 Fryers are fitted with a working thermostat for user temperature control and a high temperature (200°C) automatic cut-out device to limit the oil temperature, should the thermostat fail.

170 The electricity supply switch and/or gas valve for the fryer should be clearly marked and positioned where it can be safely operated to turn off the heat source if the oil or fat catches fire.

171 When you drain the fryer, first isolate it and allow the oil or fat to cool sufficiently before draining into a suitable container. Make sure that the inside surface of the container is dry before oil is run into it. Do not leave containers of oil in passageways or any other place where they might be knocked over. Check that the drain valve is closed before re-filling.

172 Do not use oil or fat at a higher temperature, or for a longer period, than the suppliers recommend. Do not mix oil and fats.

173 Before you switch on the heating element or light the gas check that the oil is filled to the oil level mark: when topping up to the recommended level, add the new oil or fat slowly. Do not top up from large containers, decant into smaller sized containers. DO NOT OVERFILL. Take care when shaking food in the basket not to let oil or fat drop onto the floor.

174 To reduce spitting and boiling, food should be dry before immersion into hot oil or fat. The basket should not be overloaded and should be

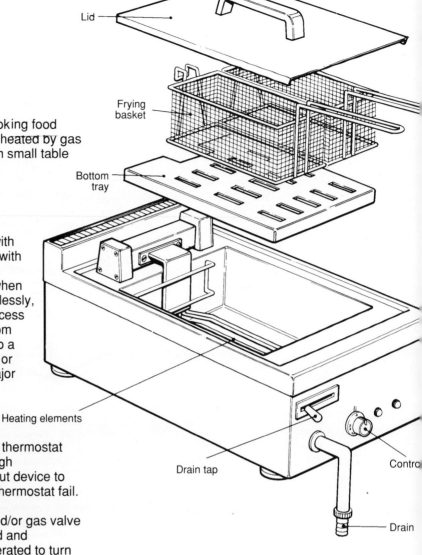

Figure 13 Table top fryer

lowered gently into the oil: do not drop it in.

175 Do not leave a working fryer unattended.

176 Appropriate fire fighting equipment and a fire blanket should be available for use in the vicinity of the fryer. A notice should be prominently displayed giving details of the action staff should take if there is a fire.

177 The local Fire Prevention Officer will give advice on the appropriate fire fighting equipment required, the procedure staff should follow in the event of a fire and the training they need to follow the procedure correctly.

Training

178 Staff should be trained to use the fryer, and to drain it safely. Training should be given in the correct procedure to follow if there is a fire, and how to use the fire fighting equipment.

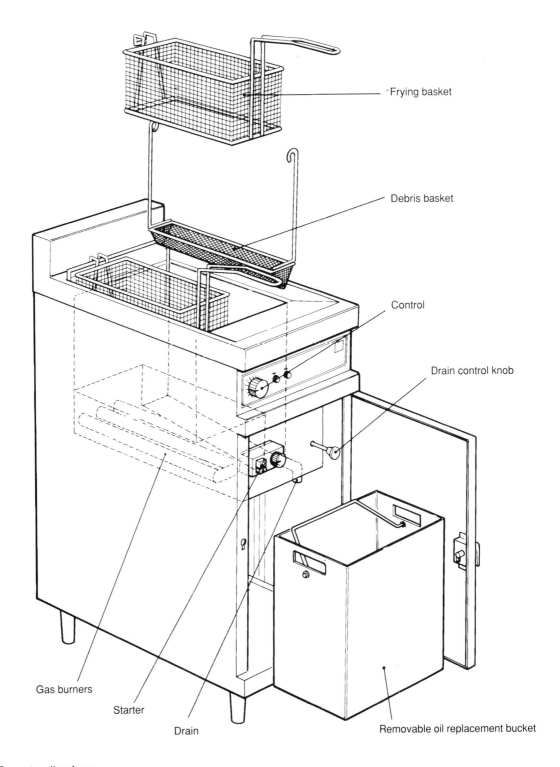

Frying basket

Debris basket

Control

Drain control knob

Gas burners

Starter

Drain

Removable oil replacement bucket

Figure 14 Free standing fryer

Cleaning

179   Before the fryer is cleaned it should be turned off and, if electric, isolated, and the oil or fat allowed to cool. Only trained staff should clean the fryer, following a laid down procedure or the manufacturer's users' handbook.

180   *Remember*
*   Check that the oil is up to the oil level mark.
*   DO NOT OVERFILL.
*   Break up dripping or fat into lumps and introduce slowly into melting fat.
*   Do not top up with oil from large containers.
*   Do not leave the fryer unattended while in use.
*   Check the food is dry before immersing in hot oil. Brush ice crystals off frozen food.
*   Do not overload the basket.
*   Do not let the basket drop into hot oil.
*   Take care when shaking food in the basket.
*   Clean up spills or drips from the floor immediately.
*   Allow the oil to cool before draining.
*   Before refilling check the drain tap is closed.

MULTI-PURPOSE COOKING PANS (BRAT PANS)

181   Multi-purpose cooking pans, or brat pans as they are more commonly called, can be used for shallow frying, stewing, braising, poaching, boiling and as a griddle. They can be power or manually tilted to pour out the contents.

Hazards

182   The main hazards associated with this type of equipment are burns from hot fat and scalds from liquids or steam.

Precautions

183   A thermostat should control the cooking temperature: the maximum recommended temperature is 190°C or 370°F.

Figure 15 Brat pan

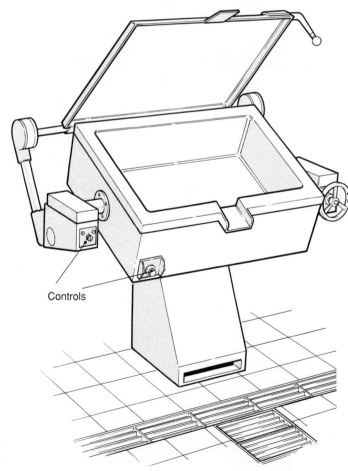

Controls

 Tilting controls

184   Never use a brat pan for deep fat frying; this is an extremely dangerous practice.

185   The tilting speed of a power tilted brat pan should not exceed 10 degrees per minute and it should stop automatically at the maximum tilt position. It should be returned to the horizontal position after tilting. Pans are designed so that they are not able to tilt accidentally.

186   If a hinged lid is fitted it should be counterweighted into the raised position. The handle of the lid should be positioned so that the operator's arms are not in the path of steam rising from the brat pan when the lid is opened. To prevent water dripping onto other equipment or the floor a lid drainage tap should be fitted to collect condensation.

187   Brat pans may have a pouring arc of up to one metre in front. Therefore to prevent spilled liquid creating a slipping hazard, suitable floor drainage channels should be provided beneath the discharge point. The manufacturer's installation handbook should provide details.

Training

188   Staff should be properly trained before being allowed to operate a tilting brat pan, with particular emphasis on the proper use of the tilting mechanism. They should be instructed never to use the pan for deep fat frying.

Cleaning

189   The brat pan should be switched off and, if electrically powered, isolated before cleaning starts. Only trained staff should clean a brat pan following laid down cleaning procedures or the instructions in the manufacturer's users' handbook.

190   *Remember*
- Never use a brat pan for deep fat frying.
- Do not stand in the way of the steam when you open the lid.
- Clean up spilled liquid immediately.

BULK BOILING PANS AND TILTING KETTLES

191   Pans and kettles are used for cooking sauces, soups, stocks, meats, vegetables and similar foods. The term applies to both large and small cooking pots installed on a worktop or the floor. They may be fixed, with a draw off tap near the bottom, or tilting, in which case the contents are poured. They may be heated by electricity, gas or steam, directly, or via a water or steam jacket.

Hazards

192   The main hazard from these appliances is scalding from steam or hot liquid splashes. The hob is often very hot and can burn if touched.

Precautions

193   To avoid the steam stand to one side when lifting the lid from a heated boiling pan. To prevent the pan boiling over, use only sufficient heat after boiling to keep food simmering, especially with soups and sauces.

194   Make sure that if a draw off tap is fitted, it is properly secured to the boiling pan.

195   All models need adequate drainage to carry away dirty water. A tundish may be sufficient for draw off models. Tilting kettles, particularly floor-mounted models, will require proper drainage channels to take account of the pouring arc during discharge.

Training

196   Staff should be trained to use pans and kettles safely, including how to remove food from them.

Cleaning

197   Only trained staff should clean a pan or kettle following a laid down procedure or the manufacturer's users' handbook.

198   Before cleaning, the pan or kettle should be turned off at the gas tap or steam valve and, if electrically powered, isolated from the supply.

199   *Remember*
- Stand to one side when lifting the lid of a heated pan or kettle.
- Do not touch the hob - it may be hot.
- Tilt the kettle carefully, do not over-tilt.
- Isolate the pan or kettle from the electricity supply and allow the it to cool before you start to clean it.
- If a drain tap is fitted and it feels loose, report it to the supervisor immediately.

Figure 16 Boiling pan / stock pot

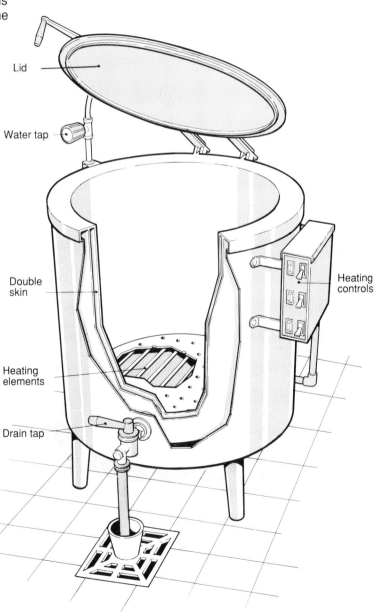

Lid

Water tap

Double skin

Heating elements

Drain tap

Heating controls

BAINS-MARIE, HOT SERVING COUNTERS AND CUPBOARDS

200   Hot food service equipment is available in a wide variety of forms; free standing modular units in the back bar style or larger floor-mounted units arranged in a number of different configurations. Some units will have heated shelves or lamps (infra-red or quartz) mounted over the serving area.

201   Bains-marie are designed for keeping cooked food hot. They may be heated by gas, electricity or steam. There are three designs:
(a)   open well: a large shallow trough containing heated water, normally found in kitchens. Pots and pans containing cooked food are placed in it to be kept hot until required for serving;
(b)   fitted container (wet type - water or steam): similar to the open well type, except that the top is constructed so that containers can be fitted into it. Cooked food is placed in these containers to keep hot: this type is normally used in a servery.
(c)   fitted container (dry type): similar to (b) except that containers are kept hot by he circulation of warm air generated within the appliance.

202   A heated cupboard may be provided under a bain-marie.

203   Hot serving counters are plain topped without any cutouts, although there may be a tiled or decorative finish, on which the food containers are placed. Heating is by conduction from the hot top, and heating lamps overhead.

Hazards

204   The main dangers from this type of equipment are burns and scalds.

Precautions

205   When removing containers or pans hold them over the trough to allow hot water to drip off. Use any lifting devices provided by the manufacturer.

206   Drain valves should discharge directly into a drain. Where this is not possible containers of adequate size should be used to collect the heating water. Take special care when carrying these containers to the drainage point.

207   Staff should wear protective gloves, to prevent them against burns from the lids of containers and the tops and sides of the unit.

208   Before the unit is used check that the heating water is up to the correct operating level in the trough and the drain tap is tightly closed.

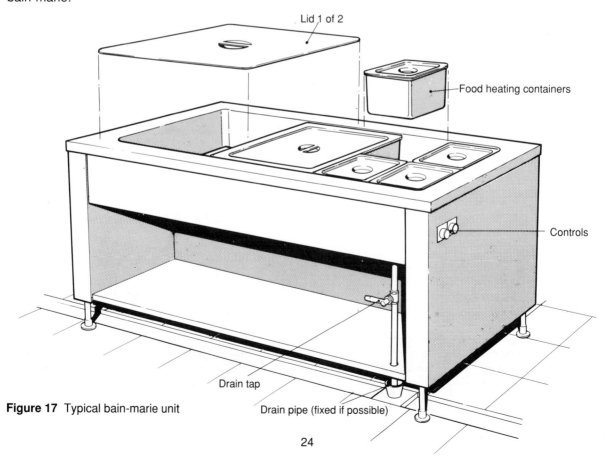

Lid 1 of 2

Food heating containers

Controls

Drain tap

Drain pipe (fixed if possible)

Figure 17 Typical bain-marie unit

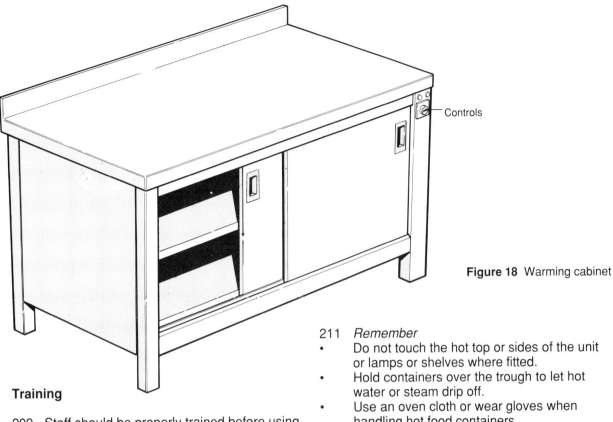

Controls

Figure 18 Warming cabinet

Training

209   Staff should be properly trained before using heated food service equipment.

Cleaning

210   Before cleaning the unit should be turned off and the water allowed to cool. Electrically heated equipment should be isolated from the supply. Only trained staff should clean the units following the manufacturer's users' handbook instructions.

211   *Remember*
*   Do not touch the hot top or sides of the unit or lamps or shelves where fitted.
*   Hold containers over the trough to let hot water or steam drip off.
*   Use an oven cloth or wear gloves when handling hot food containers.
*   Drain the heating water into suitable containers and carry them carefully.
*   Do not leave serving utensils projecting over the edge of the food containers.
*   Turn off the heat source when serving is completed.

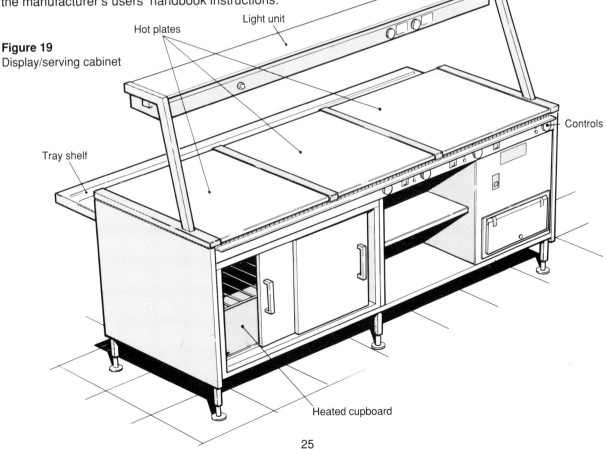

Light unit

Hot plates

Figure 19
Display/serving cabinet

Controls

Tray shelf

Heated cupboard

WATER BOILERS/CAFÉ SETS

212   While we cannot cover in detail the entire range of equipment now available, the guidance given is generally applicable.

213   Water boilers provide hot water or steam, usually for making hot drinks. The hot water and steam are provided from different outlets, the steam being injected into a cold liquid to heat it. Water boilers can be heated by gas, electricity or steam.

214   They are usually fixed to a servery with the boiler either mounted on the work surface or under the counter. There are three types:
(a)   bulk water boilers - basically a large kettle under no pressure;
(b)   expansion boilers - provide boiling water at no pressure;
(c)   pressure boilers - work at low pressure and provide boiling water and steam.

215   Free standing beverage units such as pour and serve coffee makers, hot chocolate and other liquid concentrate appliances are also available.

Hazards

216   The main hazards associated with this equipment are burns and scalds.

Precautions

217   A pressure boiler should be fitted with the following safety devices: safety valve, pressure gauge, water level gauge; if electrically heated, a low water level cut-out device; if gas heated, a flame failure cut-out device.

218   A pressure boiler and the safety devices should be inspected by a competent person, at least once every twelve months.

219   The unit should be positioned where it can be easily operated, on a fireproof base where necessary, and, for gas fired boilers, where adequate and safe flues can be installed.

220   Before the boiler is heated the cold water supply should be turned fully on. Where there is a feed water tank the water in it should be at the correct level. Staff should not interfere with or alter the water supply or heating control settings.

221   To minimise the risk of boiling water splashing, particularly from pressure boilers, place the receiving vessel right up under the

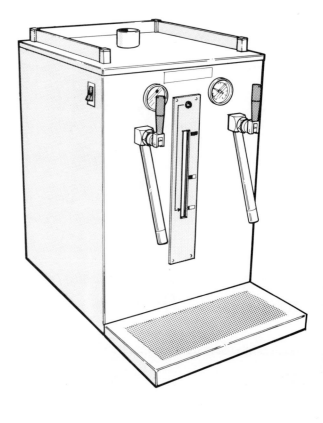

Figure 20 Table top pressurised water heater

draw-off tap. Keep a drip tray under the draw-off tap at all times.

222   If overfilled, a bulk water boiler can boil over. If this happens staff should take special care when switching it off to avoid scalding themselves.

Training

223   Staff should receive proper training in the safe use of water boilers and beverage machines.

Cleaning

224   Before starting to clean any of this equipment it should be turned off at the gas tap or steam valve or, if electric, isolated. Only trained staff should descale the equipment following the manufacturer's instructions for descaling in the users' handbook. Staff should never remove safety fittings on pressure boilers.

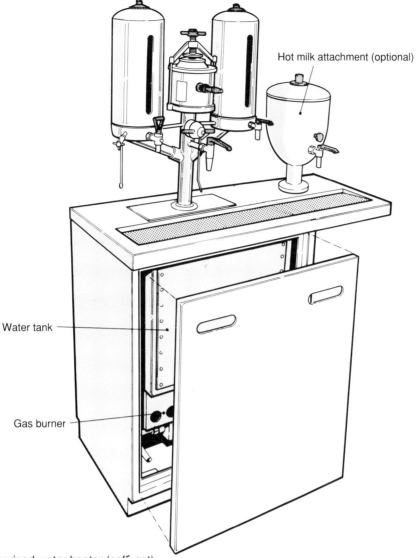

Hot milk attachment (optional)

Water tank

Gas burner

Figure 21 Pressurised water heater (café set)

225 *Remember*
- Make sure the cold water supply is fully on before you light or switch on the equipment.
- Do not alter the heating control settings on automatic units.
- Keep the pressure gauge and safety devices clean.
- Keep the drip tray in position.
- Keep the receiving vessel up to the tap to stop splashing.
- Turn off and, if electrically heated, isolate the boiler before cleaning.

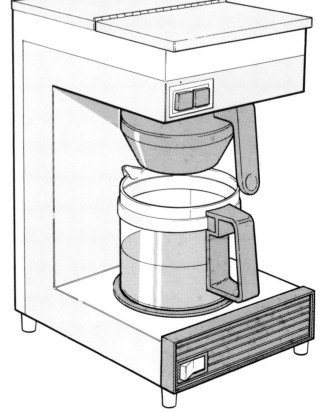

Figure 22 Pour and serve unit

URNS

226   Urns provide hot or boiling liquids such as water, milk or coffee in limited quantities. They are fitted with taps near the bottom to draw off the contents. Portable urns, and jacketed urns which have separate containers immersed in water for heating beverages, are also available. They may be heated by electricity, gas or steam.

Hazards

227   The main hazards associated with the use of urns are scalds and burns.

Precautions

228   Portable urns should be positioned on a firm level surface where they cannot be knocked against or knocked over.

229   Keep the draw-off tap clean, free from sediment and properly secured to the urn.

230   Do not let an urn boil over: as well as the possibility of scalds or burns electrical equipment can be damaged or gas flames extinguished.

231   Make sure the water level in the jacket of a jacketed urn is up to the correct level.

232   Do not let the urn boil dry.

233   Do not tilt an urn to draw off hot liquid from below the level of the tap.

234   Place the receiving vessel as close as possible to the tap to minimise splashing.

235   Keep a suitable drip tray in position under the tap.

Training

236   Staff should be trained to use urns safely.

Cleaning

237   Only trained staff should clean urns. Before cleaning the urn switch it off and, if it is electric, isolate from the supply. Make sure the urn is cool before you tilt it to remove residual cleaning water.

238   *Remember*
- Never heat other liquids in a hot water urn.
- Do not tilt an urn to draw off hot liquid from below the level of the tap.
- Do not let an urn boil dry.
- Do not let an urn boil over.

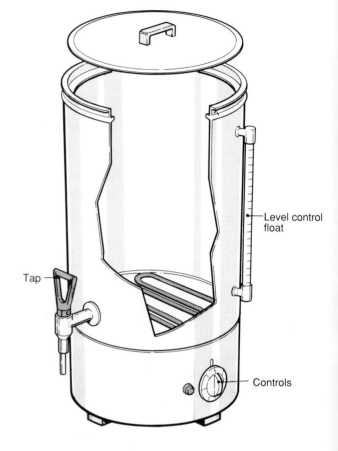

Figure 23 Water urn

- Always use the shortest possible pour to minimise splashing.
- Keep the drip tray in position.
- Make sure the water level in the jacket is up to the mark.

FLAMBÉ LAMPS

239   Flambé lamps are used in restaurants to cook food at the customer's table. They may be fuelled by methylated spirits (meths) or by butane gas from a disposable cartridge.

Hazards

240   The main hazard associated with flambé lamps is one of fire or explosion during refuelling (for meths fuelled lamp), or while changing the cartridge.

Precautions

241   Do not refuel or change a cartridge indoors. Keep a fuelled spare flambé lamp available to avoid having to refill, or change a butane cartridge, in the dark (at night in the open air) or at a busy

28

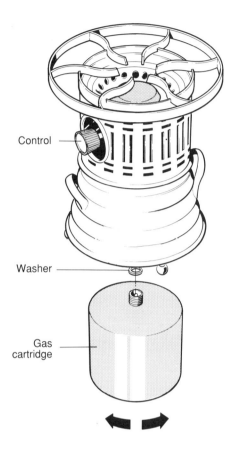

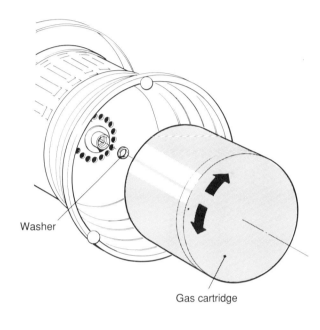

Figure 24 Flambés

Figure 25 Flambés

period. Do not refuel or change a cartridge in a public place, such as a restaurant or bar.

242 Change a cartridge only in a well ventilated space. Even in the open air there should be good air circulation.

243 Do not refuel or change a cartridge in the kitchen or restaurant. Store meths and spare butane cartridges in a safe place away from the kitchen or dining area.

Butane fuelled lamps

244 Examine the washer in the adaptor of a butane fuelled lamp regularly for wear, and replace it immediately if it becomes cracked or deformed.

245 Hold the lamp upright when changing the cartridge and do not overtighten the cartridge when screwing it into the adaptor in the lamp head.

Meths fuelled lamps

246 Allow the wick of a meths fuelled lamp to cool before the container is refilled. Take care not to spill any meths. Wipe up any spillage

immediately and carefully dispose of the cloth or paper.

Training

247 Staff should be properly trained to light and use flambé lamps. Only trained staff should refill a flambé lamp, or change the butane cartridge. Staff should know of the dangerous nature of butane gas and how important it is to follow the precautions.

248 *Remember*
• Extinguish any source of ignition in the vicinity before refuelling a flambé lamp. Do not smoke.
• Do not refill a flambé lamp in a restaurant or kitchen.
• Do not change a butane cartridge in the restaurant or kitchen.
• Always refill, or change cartridges, in the open air.
• Keep the flambé lamp upright when changing the cartridge.
• Do not overtighten the butane cartridge.
• Let the lamp cool before refuelling with meths.
• Store spare butane cartridges and meths in a safe place.

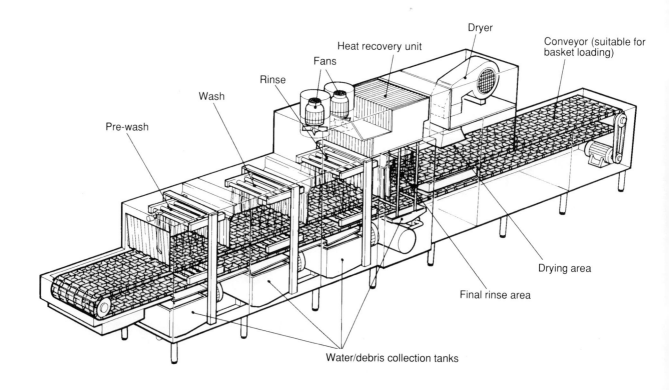

Figure 26 Large conveyor type dishwasher unit

Labels on figure:
- Dryer
- Heat recovery unit
- Conveyor (suitable for basket loading)
- Fans
- Rinse
- Wash
- Pre-wash
- Drying area
- Final rinse area
- Water/debris collection tanks

DISHWASHING MACHINES

249 Dishwashing machines are used to wash, rinse and sterilise soiled crockery and cutlery in large quantities. Some small items of kitchen equipment, such as chopping boards, mixer parts and parts of gravity feed slicers, can also be cleaned by passing through a dishwashing machine. Dishwashers may be of a batch or conveyor type. Pot washing machines are similar to dishwashing machines but as the term indicates they are used to wash saucepans, pots, trays, etc.

Hazards

250 Using a pot or dishwasher can create a number of hazards. Most types use very hot water which will scald. In some machines the water heating elements inside the dishwasher are within reach and capable of burning anyone who touches them.

251 Moving parts, such as the conveyor, can cause serious injuries.

252 Broken crockery and glass can cause cuts.

253 Because dishwashers use large amounts of water extra care is needed to make sure that no water or moisture gets into the electrical circuitry.

254 Some chemical detergents are hazardous if not used and stored properly.

255 On some machines the side panels enclosing the final rinse can get very hot and may burn anyone touching or brushing against them.

Precautions

256 Pot and dishwashers should be installed, maintained and operated only in accordance with the manufacturer's instructions. Whenever they could come into contact with the very hot water operators should wear rubber gloves and if necessary aprons and boots.

257 The conveyor and any other moving part which could cause injury should be properly guarded. Side panels should not be removed while the dishwasher is working.

258 No attempt should be made to remove broken crockery or glass while the dishwasher is working. Staff should always wear protective gloves when removing broken crockery and glass.

259 Chemical detergents can burn the skin and eyes. If chemical detergents are used the suppliers' instructions on precautions during handling should be followed. Suitable protective

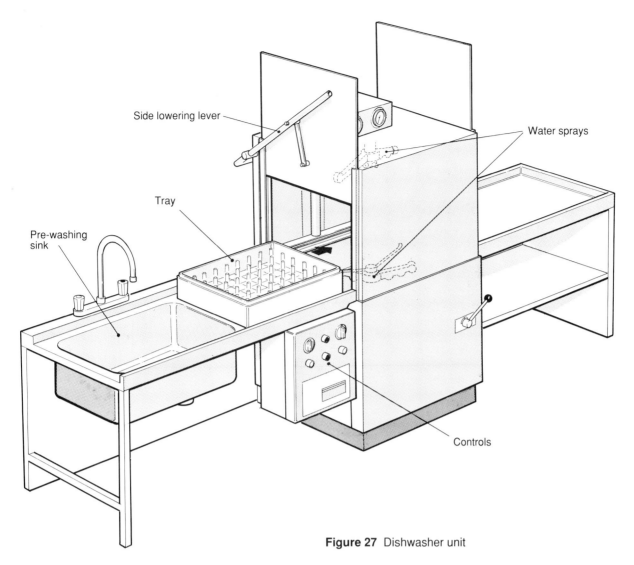

Side lowering lever

Tray

Pre-washing sink

Water sprays

Controls

Figure 27 Dishwasher unit

Figure 28 Detail of dishwasher rotating washer jet unit

clothing should always be worn. The correct strength of detergent should be used and automatic dosing equipment should be checked regularly.

Training

260   Staff must be properly trained to use pot and dishwashing machines. They should be aware of the hazards arising from the chemical detergents used, the function of the dishwasher and the safe systems of work which need to be followed for removing broken crockery and glasses etc, and for cleaning the dishwasher. Only an authorised and trained member of staff should adjust any part of the dishwasher.

Cleaning

261Before the dishwasher is cleaned it should be turned off and isolated from the electrical supply. Only a trained member of staff should clean the dishwasher following a laid down safe system of work or manufacturer's users' handbook. Under no circumstances should a conveyor dishwasher be cleaned while the conveyor is running.

262   *Remember*
• Always wear rubber gloves when working with very hot water.

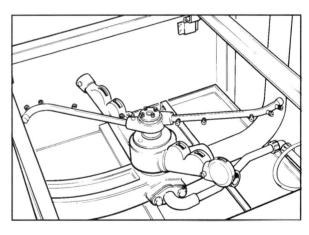

• Make sure all guards provided are in position before you start the dishwasher.
• Never put your hands in the dishwasher while it is running.
• Never try to adjust or repair any part of the dishwasher unless you have been trained to do so.
• Never let the water level go over the maximum for the dishwasher.
• Concentrated detergent can burn skin and eyes. Always wear the protective clothing provided when handling, diluting or cleaning up spilled detergent.
• The side panels by the final rinse may be very hot.

31

HEATED RINSING SINKS

263 Heated rinsing sinks contain water which is heated to a temperature in excess of 85°C or 180°F to provide a hot rinse after a detergent wash when washing up crockery, cutlery and kitchen utensils. The items being washed are often placed in baskets, then immersed in the hot clean rinsing water for a few minutes to remove final residues of food and detergent.

264 This type of sink is normally made of metal, either stainless steel or galvanised mild steel, and should be heat insulated. It may be heated by electricity or gas and the water temperature is controlled by a thermostat. Some models incorporate rotating brushes in the wash sink to scrub plates.

Hazards

265 The main hazard associated with heated rinsing sinks is scalding by very hot water. The water in the sink is always hot enough to scald if hands are immersed in the water or if the water overflows the edge of the sink.

266 Sinks fitted with scrubbing brushes can break plates if care is not taken. Deep cuts can be caused by broken crockery. If breakages occur, the sink should be drained, cleared and then refilled.

267 The floor around a sink is often wet and slippery.

268 If water or moisture gets into the electrical equipment, there is an increased risk of electric shock.

Precautions

269 Staff should be provided with suitable gloves, aprons and boots and should always wear them when working at a heated sink. The gloves should protect the forearms.

270 Long handled baskets should be used to immerse crockery etc into the hot water.

271 The maximum water level should be clearly marked and never exceeded. This reduces the risk of scalding water surging and overflowing the sink when, for example, a tray of crockery is immersed in the water.

272 Floor drains should be provided.

273 The thermostat switch should be clearly

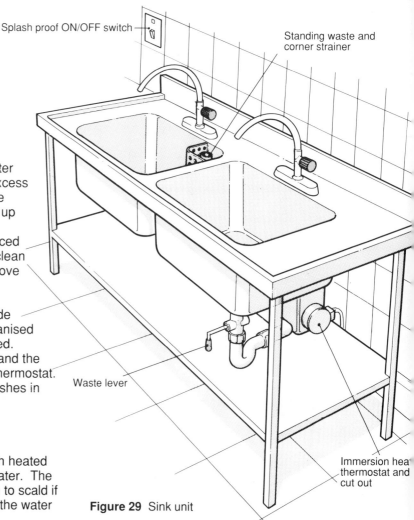

Splash proof ON/OFF switch

Standing waste and corner strainer

Waste lever

Immersion heat thermostat and cut out

Figure 29 Sink unit

marked to show if an adjustment is either increasing or decreasing the water temperature. Only trained staff should adjust the thermostat.

274 Make sure water does not get into the electrical equipment.

Training

275 All staff who work at heated rinsing sinks should be properly trained. They should know of the risk of scalding, and understand the need to wear the protective clothing provided for their use.

276 *Remember*
- Always put on your rubber gloves and other protective clothing provided before starting work at a heated rinsing sink.
- Never fill the sink above the maximum water level mark.
- Always lower baskets slowly into the water.

WASTE DISPOSAL UNITS

277 Waste disposal units dispose of food waste by mechanically grinding it up and then flushing it into a waste drain or other suitable outlet. They are usually fixed installations, either built in or free-standing units.

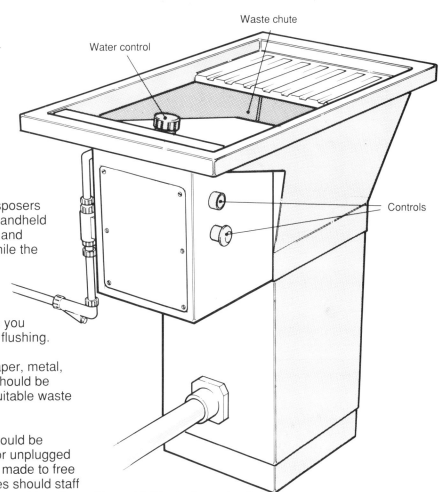

Figure 30 Waste disposal unit

Hazards

278 The main hazard from waste disposers is possibly severe injury if a hand or handheld implement is put into the feed hopper and comes into contact with the cutters while the machine is switched on.

Precautions

279 Turn on the water supply before you switch on the unit, to ensure effective flushing.

280 Unsuitable materials such as paper, metal, plastic etc and in most cases bones should be removed from the waste. Feed the suitable waste into the unit at an even rate.

281 If the unit does get blocked it should be switched off and electrically isolated or unplugged from the supply before any attempt is made to free the blockage. Under no circumstances should staff put their hands or any handheld implement into the hopper while the unit is switched on, even if it has stopped moving.

282 Guards or restrictor plates which prevent fingers reaching the cutters should always be in position before the machine is used.

Training

283 Staff who have to use a waste disposer should be trained. They should know how blockages are caused and how to avoid them, and the laid down procedure for clearing them. Staff should be warned of the risk of severe injury if they put their hands or a hand-held implement into the hopper.

Cleaning

284 Before cleaning a waste disposer should always be switched off and isolated. Only trained staff should clean it, following the laid down cleaning procedure or the instructions in the manufacturer's users' handbook.

285 *Remember*
- Remove all unsuitable material, such as plastic and metal, from waste.
- Switch on the water supply and the disposer before feeding waste in.
- Feed the waste at an even rate.
- Always isolate or unplug the machine from the power supply if it becomes blocked.

- Never, even if the machine has stopped, put your hand or any implement you are holding inside the feed hopper while the machine is switched on.

Figure 31 Small in-sink waste disposal unit

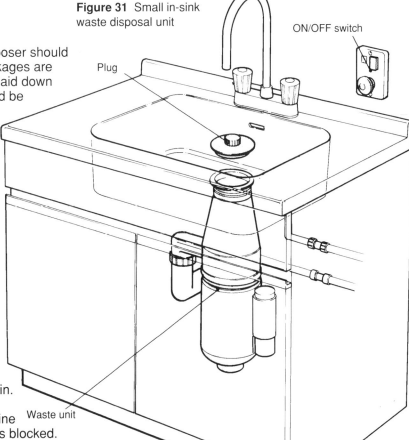

WASTE COMPACTORS

286 These machines are widely used to compress refuse into plastic bags for disposal. They can be large, high headroom, free standing, or smaller cabinet style fitted under counters and worktops.

Hazards

287 The main hazards arising from the use of high headroom machines are:
(a) injury to fingers and hands from the trap created by the moving compactor plate and the rim of the compacting chamber;
(b) injuries to the toes and feet from the trap created between the descending chamber and the base plate.

288 Glass in the bags can cause cuts to the operator's hands after compaction.

Precautions

289 On high headroom models safety limit switches should be fitted to prevent the compactor plate descending before it is in its correct position over the chamber. An interlocked guard should be fitted at the front edge of the compactor unit which should push the operator's fingers and hands away from the edge of the chamber as it moves

into the operating position. It should also prevent the operator's fingers reaching the edge of the chamber when the compacting plate is in position. These safety devices are fitted by the manufacturer. The compactor should not be used if they are not working properly.

290 The movement of the compacting chamber downwards should be controlled by a 'dead man's' control which automatically stops the movement of the chamber as soon as the operator releases the control. The last few inches of downward travel of the chamber should be non-powered.

291 Under counter cabinet type compactors should have safety switches fitted which prevent the compression action starting unless the bin or trolley is fully in position, and stop and reverse the compression action if the bin or trolley is removed during the compaction cycle. The compactor should not be used if these safety switches are not working properly.

292 Glass should not be put unprotected into the compactor. Put small glass articles in a stout cardboard box before putting them into the compactor. If large pieces or significant quantities of glass refuse are created a dedicated glass crusher or special accessory for crushing glass available from the manufacturer should be used. Special glass and tear proof bags are available.

Figure 32 Small compactor unit

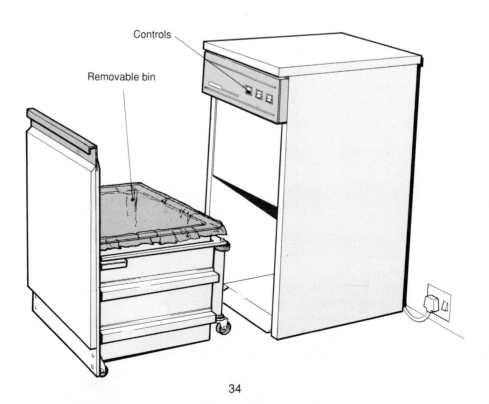

293   Operators should wear suitable protective gloves when handling waste, particularly if there is any likelihood of broken glass or other sharp or pointed waste being present.

Training

294   Staff should be properly trained before using this type of equipment. They should know what the functions of the controls and guards are and what to do if they do not work properly.

Figure 33 Compactor in closed position. Raised for the removal of waste

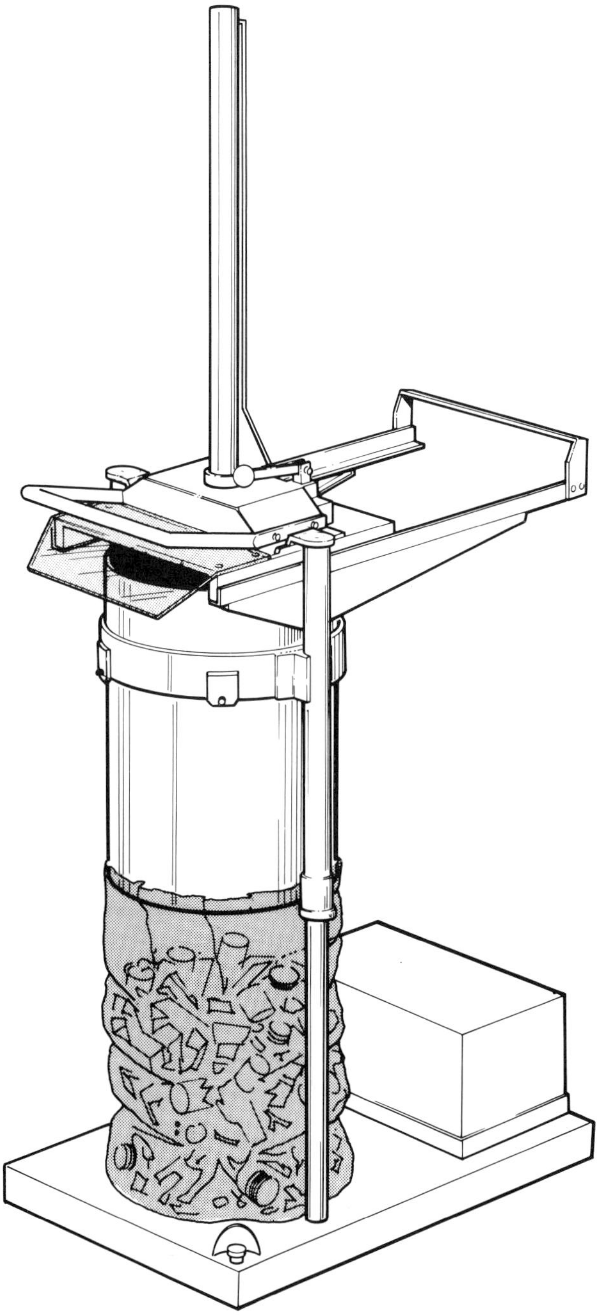

Cleaning

295   Only trained staff should clean this type of machine following a laid down cleaning procedure.

296   *Remember*
- Do not use the compactor if the guard is broken or not fitted.
- Do not use the compactor if the safety switches do not work.
- Report to the supervisor immediately if any of the controls or safety devices do not work properly.
- Do not put loose glass bottles, jars or other sharp or pointed waste objects into the compactor unless suitably wrapped.
- Always wear protective gloves when handling bagged waste.

Figure 34 Compactor in open position

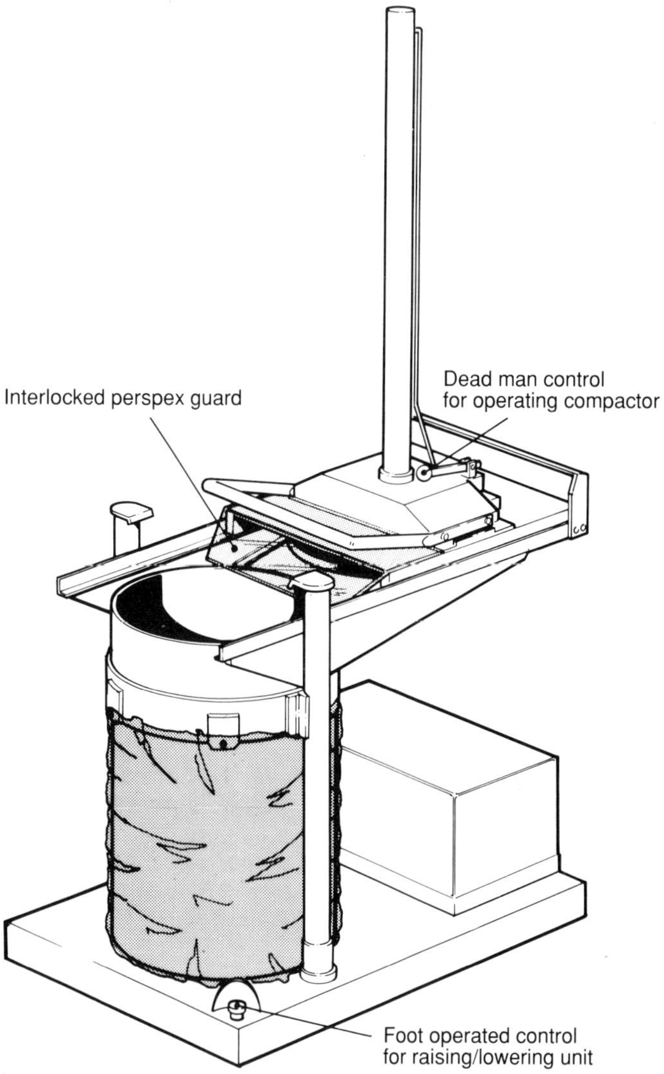

Interlocked perspex guard

Dead man control for operating compactor

Foot operated control for raising/lowering unit

CHILL SAFETY

297 This section deals with small and medium sized walk-in cold stores and chillers. A chill room normally operates at or just above freezing point and the cold store operates below freezing.

Hazards

298 The main hazard associated with chillers and cold stores is exposure to low temperatures, either while working or if accidentally locked in. Additional dangers may arise if the refrigerant escapes into the atmosphere.

299 Working in the cold store can slow down mental reactions and can reduce manual dexterity, which increases the potential for accidents.

Precautions against the effects of cold

300 Employees should not be exposed to low temperatures during the working day for any longer than absolutely necessary. Visits of less than a few minutes may not require protective clothing, but if more than one visit is made within a short time protective clothing may need to be worn. The type of protective clothing will depend on many factors, including the temperature of the cold store, type of work being carried on, the work rate and individual preference. It is particularly important to keep the hands, head and feet warm. Manufacturers/suppliers should be able to recommend suitable protective clothing.

301 Employees should be given advice on the effects of working in low temperatures and be trained to recognise early signs of frostbite (eg white patches on the skin).

First aid

302 Anyone suffering from the effects of cold should be placed in a warm room and given warm drinks. Rapid heating must be avoided. Areas of the skin showing signs of frostbite should not be massaged. Medical advice should be sought.

Precautions against being locked in

303 At least one exit from the chill or cold store should be capable of being opened from the inside, or there should be an alternative means of escape, such as a hatch. A hatch may be fitted into the door or walls and retained by a bolt or bar accessible only from the inside.

304 Doors and escape routes should not be obstructed either from inside or outside the store.

305 The opening mechanism on the escape route should be marked, eg with emergency lighting or luminous signs, so that it is clearly visible to anyone trapped inside. Opening mechanisms should be kept in good working order at all times. Instructions on the action to be taken by someone locked in should be clearly displayed inside the chill or cold room.

306 If it is necessary to lock the chill or cold store, say for security reasons, all staff should follow a system of work to ensure that the cold store is unoccupied before it is locked, and an alarm should be provided in the store to sound in a normally occupied area, or there should be an axe inside the store for breaking out.

Precautions against the accidental release of refrigerants

307 To minimise the possibility of refrigerant release plant should be designed, installed and regularly maintained by competent engineers.

308 The most common refrigerants, R12, R22 and R502, are chlorofluorocarbons. High concentrations of these refrigerants in enclosed or non-ventilated areas can lead to oxygen deficiency and asphyxiation.

309 Some older units may use ammonia as the refrigerant. Ammonia gas is both toxic and explosive.

310 If there is a major release of refrigerant the premises should be evacuated immediately and, in the case of ammonia, any naked flame in the vicinity of the plant should be extinguished.

Other safety precautions

311 Chills and cold stores should be well lit. A minimum illumination level of 300 lux is recommended. Walls, floors and doors seals should be kept free from ice build-up.

312 Racking should be appropriately designed for the goods being stored and the safe working load should be displayed. Racking systems should be inspected regularly for damage.

MACHINE SAFETY

Guarding

313 Dangerous parts of any machine must be guarded. Catering machinery has to be stripped down for cleaning more frequently than most other

machinery. It should be possible to clean guards easily and thoroughly and they must be replaced after cleaning. Machines must not be run if any guard has been removed.

314   Guards should be designed and made only by someone who understands the principles and standards involved.

Drives

315   Drives and transmission machinery must be enclosed by a guard or safely situated within the machine body.

Feed and delivery openings

316   Many machines have openings to allow raw materials to be fed in or the finished product to be taken out. The openings must not allow anyone to reach into the dangerous parts of the machine.

Fixed guards

317   Fixed guards must be secure and tamperproof and their fixings should be removeable only with a tool, eg bolts or hexagon socket machine screws. Toggle clamps, wing nuts and quick release catches should not be used.

Electrical interlocking

318   Guards that have to be opened regularly are best fitted with interlocking switches connected to the motor supply, so that the machine cannot start or run unless the guard is in place. It is essential that any switch used is of proven reliability and installed in such a way that it will not be prone to failure or misuse. (Ordinary reed switches and micro switches are not normally suitable.)

Maintenance of guards

319   Whatever the type of guarding and interlocking, it should be checked before the machine is used, and maintained in proper working order. A visual examination should be made and any broken or missing guards should be repaired or replaced. Interlocks should also be checked and tested to ensure they are secure and working. Broken interlocks should be replaced or repaired.

320   In particular, the guards should be checked after maintenance or cleaning when they may have been removed.

Machine setting

321   Some machines have to be adjusted while running. Final settings sometimes have to be made once the actual product can be seen. The controls for running adjustments should be safely positioned. Machines should be set and adjustments made with the guards in position.

Machine stability

322   Machines should be on a secure base so that they cannot move or vibrate when in use. They may need to be bolted to the floor or worktop.

Operator safety

323   Machine operators should not wear loose or frayed clothing, or jewellery.

324   Machines should not be used if the operator is feeling unwell or drowsy (certain medicines etc carry a warning that they may cause drowsiness).

Warning notices

325   Warning notices may be displayed alongside machines to remind operators and others of the dangers they pose. Many machine suppliers provide suitable notices, eg:

> # Dangerous Machine
> ## to be operated by authorised persons only
>
> ## *Warning*
> ## Do not talk or distract the attention of the operator while the machine is in motion

No-volt releases

326   New machines with exposed blades, such as slicers, are fitted with a no-volt release (NVR). This device ensures that the machine starts only when the control button is operated and not when it is plugged in or when the electrical power is switched on.

327   If an existing machine is to have a major overhaul it should, if possible, be fitted with a no-volt release at the same time. Consult the manufacturers for advice.

SLICERS

328   There are three main types of slicer: the gravity feed slicer, the horizontal feed slicer and the bacon slicer.

329   Gravity feed slicers have an inclined carriage and are used for slicing cooked meat and other foods. The carriage may be hand or power operated.

330   Horizontal feed slicers have a horizontal carriage to carry the meat. The operator pushes this carriage towards a circular blade and slice thickness plate.

331   The traditional bacon slicer has a vertical circular blade and a horizontal carriage. The carriage has clamps to hold the meat. Some machines are power operated and present additional hazards. These machines should not be used to slice slippery, small or unevenly shaped food, eg cucumbers or tomatoes, that cannot be securely held or clamped to the carriage.

Hazards

332   The main hazard on these machines is the exposed cutting edge of the blade, which can cause serious cuts and even amputation.

333   On power driven gravity feed slicers there may also be a possibility of the operator getting trapped between the carriage and the machine frame.

334   On power driven bacon slicers it is possible for the operator's hands to be pushed onto the blade by the meat supports on the carriage.

Precautions

Gravity and horizontal feed slicers

335   The edge of the blade must be guarded (blade guard) except at the cutting section. The guard may be permanently fixed to the machine frame flush with the blade edge or it may be removed for cleaning if it interlocks the power supply to the blade motor.

336   A thumb guard should be provided at the operator's side of the carriage to cover the blade at the far end of each cut.

337   On certain machines the blades have to be removed for cleaning. A blade carrier that prevents access to the sharp edge should be used to remove the blade safely.

338   The slice thickness plate should be shaped to the edge of the blade to prevent injury at the cutting section.

339   The carriage should have a last slice device or meat pusher of suitable size and shape to prevent the operator's hand slipping onto the blade. The meat pusher should have a handle and on the horizontal slicer it should not be possible to swing the pusher clear of the carriage.

340   Keep the blade sharp. The operator has to use more force with a blunt blade which increases the risk of slipping onto it.

341   When the blade sharpener is in use all guards must be in place. The slice thickness plate should be at zero except on machines with a detachable sharpener.

342   On gravity feed slicers a suitable plastic or perforated metal carriage guard should be fitted at the operator's side of the carriage to prevent accidental contact with the blade.

Figure 35 Gravity feed slicer

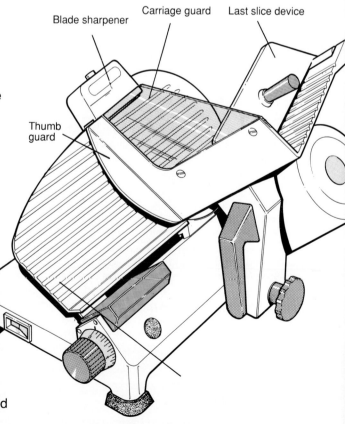

FOOD PROCESSORS

343   Food processors consist of a motor drive base to which various bowls, cutter plates or mixing blades can be fitted. They can be used for slicing, grating, mixing and liquidising.

Hazards

344   The cutter plates and mixing blade rotate at high speeds and can cause severe injury to the fingers. Also, hot ingredients may be ejected from the bowl and cause scalding.

Figure 36 Slicing, dicing , grating and chipping device

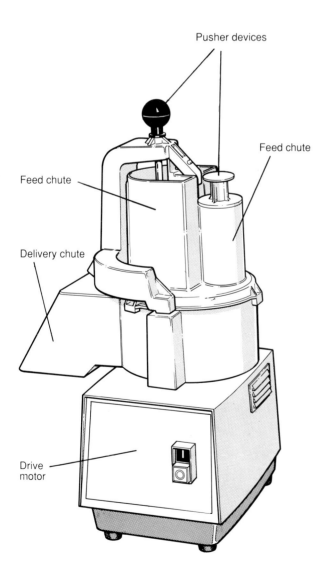

Pusher devices

Feed chute

Feed chute

Delivery chute

Drive motor

Precautions

Slicing, dicing, grating and chipping devices

345   The feed chutes and pusher devices should prevent access to the cutter when it is running. Pushers should be connected to the machine by a fixed linkage. Loose pushers tend to get lost. Separate cucumber slots may have loose pushers if finger access to the cutter is not possible.

346   Machines are usually designed so that cutter plates can be changed quickly. There is a risk of injury if the cutter can be exposed when it is moving. The blade cover should interlock the power supply to the motor. A braked motor or a time-delay interlock may be needed to prevent access to the cutter before it has stopped.

347   Access to the underside of the moving cutter must be prevented. This may be achieved by an elbow in the delivery chute or a smooth plate below the cutter plate which throws the cut material into the delivery chute.

Mixing and liquidising bowls

348   Mixing and liquidising bowls must be fitted with a lid to prevent material being ejected or fingers reaching to the moving mixing blades. The lid should interlock the supply to the drive motor. If a feed chute is fitted so that ingredients can be added during mixing, it should be designed to prevent fingers touching the blades.

Existing machinery

349   Many existing machines do not fully satisfy the above standards. If they are to continue in use they should be maintained to at least their original new condition. It is strongly recommended that any suitable modifications available from the manufacturer should be obtained and used.

Training

350   Food processors are prescribed dangerous machines. Only a fully trained operator or a trainee under close supervision may use them. The training must make clear the risks and how to avoid them.

Cleaning

351   Before you start to clean the machine, unplug it or switch it off at the isolator. Do not rely on machine controls or interlock switches.

PLANETARY MIXERS

352 This widely used mixer has a single beater which moves around the stationary bowl so that material at the sides is thoroughly mixed. The beater may be a metal grid, a hook or a whisk. Many machines also have other attachments such as vegetable slicers connected to them.

Hazards

353 Hands can be bruised and even crushed in the trap between the beater and the bowl. There is a risk of this happening if the operator adds ingredients or scrapes down the bowl during mixing. The larger the machine the greater the hazard.

354 Some machines have gear or clutch levers which can easily be knocked into engagement. This is more likely on old or poorly maintained machines with worn mechanisms. Most modern machines have electronic speed controls or gear box designs that do not have this problem.

Precautions

355 Recipes and working methods should be used that avoid operators putting their hands in the bowl.

356 Bowl extension rings should be used whenever possible to restrict access to the beater.

357 On larger machines a full mixing bowl can be very heavy. Castor-framed trolleys may be used to prevent strain injuries.

358 Ensure that the gear or clutch lever cannot easily be knocked into the operating position. The machine should be regularly serviced, preferably by the manufacturer or the manufacturer's agent.

359 Suitable warning notices should be posted beside the machine.

Training

360 The operator should be trained to work the machine safely. The training should make clear the risks and how to avoid them.

Cleaning

361 Only a trained person should clean the machine. Consult the instruction manual for cleaning procedures.

362 Before you start to clean the machine, unplug it or switch it off at the isolator. Do not rely on the machine controls.

363 *Remember*
- Do not try to use or clean the machine without proper training.
- When the machine is running -
  never try to feel the mix;
  never try to scrape down the bowl;
  never reach into the bowl when adding ingredients.
- Use a bowl extension ring whenever possible.
- Make sure the gear or clutch lever cannot fall or be knocked into gear.
- Always unplug the machine or switch it off at the isolator before you start to clean it.

Figure 37 Planetary mixer

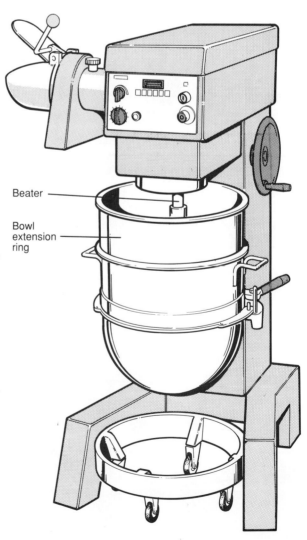

Beater

Bowl extension ring

Advice on other machines used in the food industry can be found in HSE booklet HS(G)35, *Catering safety: food preparation machinery.*

FURTHER INFORMATION

HSE priced publications (available from HMSO)

Legal guidance

HS(R)4 (rev) *Guide to the OSRP Act 1963* 1989
ISBN 0 11 885463 1

HS(R)6 (rev) *Guide to the HSW Act* 1990
ISBN 0 11 885502 6

HS(R)23 *Guide to the Reporting of Injuries, Diseases and Dangerous Occurrences Regulations 1985* 1986 ISBN 0 11 883858 X

HS(R)25 *Memorandum of guidance on the Electricity at Work Regulations 1989* 1989
ISBN 0 11 883963 2

Approved Codes of Practice

COP 1 *Safety representatives and safety committees* 1988 ISBN 0 11 883959 4

COP 20 *Standards of training in safe gas installation* 1989 ISBN 0 11 883966 7

COP 29 *Control of Substances Hazardous to Health Regulations 1988* 1988 ISBN 0 11 885468 2

COP 42 *First aid at work* 1990 ISBN 0 11 885536 0

Gas safety

HS(G) 34 *Storage of LPG at fixed installations* 1987
ISBN 0 11 883908 X

CS 4 *Keeping of LPG in cylinders and similar containers* 1986 ISBN 0 11 883539 4

Electrical safety

GS 27 *Protection against electric shock* 1984
ISBN 0 11 883583 1

GS 37 *Flexible leads, plugs, sockets and other accessories* 1985 ISBN 0 11 883519 X

PM 29(rev) *Electrical hazards from steam/water pressure cleaners etc* 1988 ISBN 0 11 883538 6
PM 32 *The safe use of portable electrical apparatus (electrical safety)* 1983
ISBN 0 11 883563 7

Food preparation

HS(G) 35 *Catering safety : food preparation machinery* 1987 ISBN 0 11 883910 1

HS(G) 45 *Safety in meat preparation : guidance for butchers* 1988 ISBN 0 11 885461 5

PM 33 *Safety of bandsaws in the food industry* 1983 ISBN 0 11 883564 5

General

Writing your health and safety policy statement 1989 ISBN 0 11 885510 7

COSHH assessments : a step by step guide to assessment and the skills needed for it 1988
ISBN 0 11 885470 4

HS(G)38 *Lighting at work* 1987
ISBN 0 11 883964 0

Watch your step : prevention of slipping, tripping and falling accidents at work 1985
ISBN 0 11 883782 6

British Standards (available from British Standards Institution, Linford Wood, Milton Keynes, MK 14 6LE)

BS 4293 *Specification for residual current operated circuit breakers* 1983

BS 6173 *Installation of gas catering appliances 1982*

BS 4343 *Specification for industrial plugs, socket outlets and couplers and for ac and dc supplies 1968*

Additional guidance

Hotel & Catering Training Board *Health and safety in hotels and catering* 2nd ed 1986
ISBN 070 330120 1. Available from: Hotel & Catering Company, International House, High Street, Ealing, London W5 5DB

General, Municipal, Boilermakers and Allied Trades Union *Risks a la carte: safety reps guide to catering hazards* 2nd ed. Available from: GMBATU, Thorne House, Ruxley Ridge, Claygate, Esher, Surrey KT10 0TL

Chartered Institution of Building Services Engineers (CIBSE) *Code for interior light* 1984 ISBN 090 095327 6. Available from CIBSE, Delta House, 222 Balham High Road, London SW12 9BS

ACKNOWLEDGEMENTS

The Health and Safety Executive gratefully acknowledges the assistance and support given by local authorities, manufacturers, institutions and organisations listed here in the preparation of this guidance.

Institution of Environmental Health Officers
Association of Metropolitan Authorities
The Royal Environmental Health Institute of Scotland
Association of District Councils
Association of County Councils
Convention of Scottish Local Authorities
Hotel Catering Institutional Management Association
The Electricity Association
Catering Equipment Users Association
Hotel Catering Training Company
Bakers Food and Allied Workers Union
Transport and General Workers Union
General Municipal Boilermakers and Allied Trades Union
Allied Breweries Ltd
Catering Equipment Manufacturers Association of Great Britain
Confederation of British Industry
Trades Union Congress
British Retailers Association
Bass UK Ltd
Scottish and Newcastle Breweries plc
J Lyons and Co Ltd
Department of Health and Social Security
The Brewers Society
British Gas plc

HSE AREA OFFICES

Area	Address	Telephone no.	Local authorities within each area
1 SOUTH WEST	Inter City House Victoria Street, Bristol BS1 6AN	0272 290681	Avon, Cornwall, Devon , Gloucestershire, Somerset, Isles of Scilly.
2 SOUTH	Priestley House, Priestley Road, Basingstoke RG24 9NW	0256 473181	Berkshire, Dorset, Hampshire Isle of Wight, Wiltshire.
3 SOUTH EAST	3 East Grinstead House, London Road, East Grinstead, West Sussex RH19 1RR	0342 26922	Kent, Surrey, East Sussex, West Sussex.
5 LONDON NORTH	Maritime House, 1 Linton Road, Barking, Essex IG11 8HF	081-594 5522	Barking and Dagenham, Barnet, Brent, Camden, Ealing, Enfield, Hackney, Haringey, Harrow, Havering, Islington, Newham, Redbridge, Tower Hamlets, Waltham Forest.
6 LONDON SOUTH	1 Long Lane, London SE1 4PG	071-407 8911	Bexley, Bromley, City of London, Croydon, Greenwich, Hammersmith and Fulham, Hillingdon, Hounslow, Kensington and Chelsea, Kingston, Lambeth, Lewisham, Merton, Richmond, Southwark, Sutton, Wandsworth, Westminster.
7 EAST ANGLIA	39 Baddow Road, Chelmsford, Essex CM2 0HL	0245 284661	Essex, except the London Boroughs in Essex covered by Area 5; Norfolk, Suffolk.
8 NORTHERN HOME COUNTIES	14 Cardiff Road, Luton, Beds LU1 1PP	0582 34121	Bedfordshire, Buckinghamshire, Cambridgeshire, Hertfordshire.
9 EAST MIDLANDS	5th Floor, Belgrave House, 1 Greyfriars, Northampton NN1 2BS	0604 21233	Leicestershire, Northamptonshire, Oxfordshire, Warwickshire.
10 WEST MIDLANDS	McLaren Building, 2 Masshouse Circus, Queensway, Birmingham B4 7NP	021-200 2299	West Midlands.
11 WALES	Brunel House, 2 Fitzalen Road, Cardiff CF2 1SH	0222 473777	Clwyd, Dyfed, Gwent, Mid Glamorgan, Powys, South Glamorgan, West Glamorgan.
12 MARCHES	The Marches House, Midway, Newcastle-under-Lyme, Staffs ST5 1DT	0782 717181	Hereford and Worcester, Salop, Staffordshire.

Area	Address	Telephone no.	Local authorities within each area
13 NORTH MIDLANDS	Birbeck House, Trinity Square, Nottingham NG1 4AU	0602 470712	Derbyshire, Lincolnshire, Nottinghamshire.
14 SOUTH YORKSHIRE & HUMBERSIDE	Sovereign House, 40 Silver Street, Sheffield S1 2ES	0742 739081	Humberside, South Yorkshire.
15 WEST & NORTH YORKSHIRE	8 St Pauls Street, Leeds LS1 2LE	0532 446191	North Yorkshire, West Yorkshire.
16 GREATER MANCHESTER	Quay House, Quay Street, Manchester M3 3JB	061 831 7111	Greater Manchester.
17 MERSEYSIDE	The Triad, Stanley Road, Bootle L30 3PG	051 922 7211	Cheshire, Merseyside.
18 NORTH WEST	Victoria House, Ormskirk Road, Preston PR1 1HH	0772 59321	Cumbria, Lancashire.
19 NORTH EAST	Arden House, Regent Centre, Gosforth, Newcastle-upon-Tyne NE3 3JN	091 284 8448	Cleveland, Durham, Northumberland, Tyne and Wear
20 SCOTLAND EAST	Belford House, 59 Belford Road, Edinburgh EH4 3UE	031 225 1313	Borders, Central, Fife, Grampian, Highland, Lothian, Tayside and the island areas of Orkney and Shetland.
21 SCOTLAND WEST	314 St Vincent Street, Glasgow G3 8XG	041 204 2646	Dumfries and Galloway, Strathclyde and the Western Isles.

Printed in the UK for HMSO   C150   1/91